John Seller

Atlas Maritimus

John Seller

Atlas Maritimus

ISBN/EAN: 9783744762397

Printed in Europe, USA, Canada, Australia, Japan

Cover: Foto ©berggeist007 / pixelio.de

More available books at **www.hansebooks.com**

Charles R.

CHARLES the Second, By the Grace of God, King of England, Scotland, France, and Ireland, Defender of the Faith, &c. To all Our loving Subjects, of what degree, condition, or quality soever, within any Our Kingdoms or Dominions, greeting: Whereas We have been given to understand, that Our Trusty and Wellbeloved Subject, John Seller, Our Hydrographer in Ordinary, hath been for these several years last past, Collecting and composing two large Treatises of Navigation, the one Entituled the English Pilot, the other the Sea Atlas, Describing the Sea-Coasts, Capes, head-lands, Bayes, Roads, Rivers, Harbours, Rocks, Sands, Soundings, Shoals, and places of Danger in most of the known parts of the World; a Work of very great Expence and Cost, and not heretofore performed in this Our Kingdom. The first Part whereof being now fully and entirely finished, We are informed that Endeavours are made by some of Our Subjects, secretly to Copy and Reprint the same, but under another Title, to the great prejudice and discouragement of the said John Seller. We therefore taking the same into Our Princely Consideration, and minding the great usefulness of this Work, have thought fit, for his future Encouragement, hereby to declare Our Pleasure, and accordingly We do by these presents strictly prohibit and forbid all Our Subjects, within Our Kingdoms of Great Britain and Ireland, to Copy, Epitomize, or Reprint the said Treatises of Navigation, (Entituled the English Pilot, and the Sea Atlas) in whole or in part, or under any other Name or Title whatsoever; Or to Copy or Counterfeit any of the Maps, Plats, or Charts that shall be in the said Treatises, within the term of thirty years next ensuing the date of these Presents, without the consent and approbation of him the said John Seller, his Heirs, Executors or Assigns: And that no such Books, Maps, Charts or Plats, or any Part or Copy thereof, be imported from beyond the Seas, either under the Name of Dutch Waggoners, or Lightning Columes, or under any other Name whatsoever, during the said term of thirty years, As the Persons offending will answer the contrary, not only by the forfeiture of the said Books, Plats, Charts, or Maps, but at their utmost peril: Whereof as well the Wardens and Company of Stationers of Our City of London; As all and singular our Officers of Our Customs in our Port of London, or any other Place within Our Dominions; And all other Our loving Subjects, whom it may concern, are to take particular notice, that due obedience be given to this Our Royal Command accordingly. Given under Our Signet and Sign Manuel, at Our Court at Whitehall, the 22th day of March, 6⅞, in the 23th year of Our Reign.

By His Majesties Command.

Arlington.

ATLAS MARITIMUS,
or A Book of
CHARTS.

Describeing the Sea Coasts Capes
Headlands Sands Shoals Rocks and Dangers.
The Bayes Roads Harbors Rivers and Ports, in
most of the knowne parts of the

WORLD.

With the true Courses and distances, from one
place to another, Gathered from the latest and
best Discoveryes, that have bin made by divers
Able and Exoerienced Navigators, of our English
Nation, Accommodated with an Hydrographicall
Description of the whole World.
By John Seller,
Hidrographer to y[e] Kings most Excellent Majestie.
And are to be Sold by him at the Hermitage Staires
in Wapping, and at his Shop in Exchainge Alley
Neer the Royall Exchange.
in London.
Cum Privilegio.

LONDON

IOHN PERCIVALLE.

THE
SEA-ATLAS:
CONTAINING
An Hydrographical Description of the SEA-COASTS *of most of the known Parts of the* WORLD.

THat whole Mass of Waters which maketh up one part of this Terrestrial Globe, and is sometimes, as it were, enriched with the Earth, as in Rivers, Streights, and smaller Seas; sometimes dasting it self into larger Floods, doth encompass the Earth, as in the Ocean or greater Seas, may be conveniently divided, somewhat according to the four general Regions or Divisions of the Earth, into four parts: The North Sea, or *Mar del Noort*, comprehendeth all those Waters which, from the Pole-Attick even unto the Equator, do wash the Shores of *Europe*, *Africa*, and *America*: The *Ethiopian Sea*, or *Mar d' Ethiopia*, which from the Equinoctial Line northerly, the Shores of *Ethiopia* on the East, and the Coasts of *America* on the West part, runneth with unknown bounds towards the Antartick Pole. The *Indian Sea*, or *Mar d' India*, bounded on the West with the Oriental Parts of *Africa*: on the North, by the South Coasts of *Asia*; and circumvironing all the Islands of the *East-Indies*, as far as *Isles de Ladrones*, and *Nova Guinea*, hath its South parts tending towards the Antartick Pole, not yet discovered. The South Sea, called also *Mar del Zur*, or *Mare Pacificum*, runneth all along the Western Shores of *America* on the one side; is contiguate with the *Indian* Sea on the other, but hath yet found no limits towards the Artick or Antartick Poles. Which general division of the Ocean, so far as conveniency may admit in the succeeding brevity Description of the Sea-Coasts, shall be observed.

The Coasts of those two famous Islands of *Great Britain* and *Ireland*, are the first that discover themselves to us, out of this Northern Division of the Ocean: The first whereof is not without cause esteemed the Metropolitan Island of *Europe*, I will say (taken in all respects) of the World; It is attended by many lesser Islands, the chief whereof are *Thanet*, *Wight*, *Silly*, *Anglesey*, *Man*, *Lewis*, the *Hebrides*, *Orkney*, *Shetland*, and *Far*: Stored with plenty of Ports, Bays, Rivers, Roads, and Harbours, capable to receive Ships of great Burthen; amongst which *London* accounted the Mart of Christendom, the Metropolis of *Great Britain*, conveniently seated on the River of *Thames*, hath the chiefest note.

Next unto which, on the East side northerly, by the *German* Ocean, are *Harwich*, *Yarmouth*, *Lin*, *Kingston* upon Hull, *New-castle*, a gallant Haven, famous for its inexhaustible Cole-Mines, and *Edenborgh* and *Dundee* in *Scotland*, &c.

On the South side, between the Coasts of *England* and *France*, called the *Channel*, are *Dover*, *Chichester*, *Portsmouth*, *Weymouth*, *Plymouth*, and *Dartmouth*.

On the West side, over against the Coasts of *Ireland*, in that violent and turbulent Sea, called St. *George* his *Channel*, are *Bristol*, *Pembrock*, or *Milford-Haven*, *Chester*, *Leverpool*, *Carlisle*, &c.

The Island (especially *England*) yeelding abundant plenty of Corn and Cattel, besides other Commodities, as Lead, Tin, Iron, Sea-cole, Saffron, Wooll, Cloth, Licorish, Mill-stones, and other rich Merchandice; multitude of Ships being continually in the Ports, serving either to export her own, or to import other Commodities from most places of the World in lieu thereof.

On the Coasts of *Ireland*, are *Knock-fargus*, *Dublin*, *Waterford*, *King-sale*, *Limrick*, *Galloway*, &c. Principally abounding in Cattel, from whence great numbers are yearly exported into other Countreys.

Passing on northerly, just under the Artick Circle, lyeth encompassed, by the Northern Ocean, or *Mare Glaciale*, *Island*, or rather *Iceland*, first discovered by one *Naddoc* a Pirat, who by a Tempest was driven to the Shores of this Countrey; which afterwards, from the coldness and store of Ice there continually found, was so named, and first inhabited by the *Norvegians*, now under the power of the King of *Denmark*; a place frequented by *Danes*, *English*, High and Low *Dutch*, and *Biscainers*, where in exchange of Bisket, Beer, Iron, Copper, Cloth, and some other Wares they being from thence, Stock-fish, and divers sorts of Fish, Train-Oyl, Skins of Foxes and other Beasts, Sulphur, and a sort of course Cloth, and Stockings, called *Wadmoll*.

The Ports most frequented, are *Strom*, *Warloswick*, *Krblewick*, *Bassand*, *Orbrack*, *Hola*, and *Haffenford*: near unto which standeth *Bellsteede*, the Residence of the Governour, a Dwelling suitable enough to the manner and fashion of this Countrey.

North-east from hence, in the Latitude of 76 and 80, lyeth *Greenland*, or King *James* his New Land, first found out by Sir *Hugh Willoughby*, in the year 1553. (although the Dutch men affirm it to be discovered first by *Jacob Heemskerk*, *William Barentson*, and *John Cornelius Rip*, Anno 1596.) which whether it be an Island or contiguate with the main Continent of *Greenland*, or some other Northern Region, none have hitherto known.

The Inland parts are stored with great numbers of Bears, Deer, Foxes, and such like Creatures; and the Sea-shores with multitude of Morses and Whales of incredible magnitude, for the catching whereof, the

(marginal notes:) *Ireland.* *Island.* *Greenland.* *Britain.*

the Inhabitants of most Sea-Ports in the Northern Ocean do usually make their yearly Voyages.

Cherry Island. Not far from hence lyeth *Bear Island*, or rather *Cherry Island*, so named from Sir *Francis Cherry* Merchant, who was at the charge of the discovery thereof; whither resort great number of Sea-Horses or Morses, and Whales; the Moscovy Company once making great profit of the trade therein.

Nova Zemla. Thirty degrees to the South-eastward hereof, is the Island of *Nova Zemla*, separated from the North Continent of *Russia*, by the Streights of *Vaigats*, alias *Fretum Naswough*, through which so many brave and worthy Navigators have attempted to find a passage

N. E. Passage. into *Scythia* and *China*; but being obstructed by the abundance of Ice met with in those Seas, could yet discover no farther eastwards then the great River *Oby*,

Tartarian Sea. the North-west confine of the Kingdom of *Tartaria*, though it hath been often reported by the *Samoid Tartars*, *Russes*, and others that have travelled those Countries by Land, that the *Tartarian* Seas do at certain seasons of the year lie open, and free from such incumbrance.

Russia. To return therefore by the known Parts of *Russia*, the Port and Places whereof, worthy observation, and most frequented, are, first *Pettzora Reca*, the Isle of *Colgoyo*, situate at the mouth of a great Bay, whence cunning by *Cape Candenoes*, there is the entrance into

White-Sea. the White-Sea, or Bay of St. *Nicholas*, which Master *Richard Chancellor*, in the *Richard-Bonaventure*, first discovered, and setled a Trade with the *Muscovites*, or *Russes*, at the Town of St. *Nicholas*, a well known Port, situate at the influx of the River *Duina*, into the Bay. But St. *Michaels*, on the Sea-side, commonly called *Arch-Angel*, is the Town of greatest Trade, especially by the *English*, who have of late there fixed their Staple.

The principal Commodities they send abroad, are Rich Furs, Hemp, Flax, Fish, Train Oyl, Pitch, Wax, Pitch, Rosin, and the like; receiving in return, Cloth, Silk, Tapestry, and some other Merchandize.

Lapland. Without this Bay, on the Coast of *Lapland*, *Finmark*, and *Norway*, are *Kola*, *Kegor*, (as at unto which Sir *Hugh Willoughby*, with his Company, in the *Bona Esperanza*, attempting first the discovery of unknown Places in this Icy Sea, were frozen to death) next are *Wardhouse*, and the *North-Cape*, so called, because it is the out-most Northern Bound of the Continent of *Europe*.

Norway. *Dronten*, in the Latin *Nidrosia*, so called, from the River *Nider*, on which it is seated; anciently the Metropolis of *Norway*, but since the subjection of this Countrey to the *Danes*, reduced to a Burrough. *Bergen* the principal Town of this Countrey, the ordinary Residence of the Governour for the Kings of *Denmark*; strongly scituate amongst high Mountains, at the bottom of a deep Creek or Arm of the *German* Ocean, called *Carmfunt*, a safe and noted Port, much resorted to by Merchants of most European Nations, bringing thither Corn, Bread, Beer, Wine, and Brandy, to supply the natural wants and defects hereof; and in exchange transporting Fish, Furrs, Boards, Cordage, Masts, and other Materials for Shipping. Then *Lanyfound*, *Anstoo*, *Marstrand*, and *Gottenburgh*, noted for the multitude of Herrings thereabout.

Not far from hence is the entrance into the *Baltick* Sea, which beginneth at the narrow Passage called the

Sound. *Sound*, and interlacing the Countries of *Denmark*, *Swedland*, *Poland*, and *Germany*, extending even to *Livonia* and *Lithuania*. The Islands whereof are many in number, the chief are, *Zealand*, *Funen*, *Langeland*,

Baltick Sea. *Laland*, *Falster*, *Alsen*, *Meon*, *Rugen*, *Bornholm*, *Oeland*, *Gottland*, *Osel*, *Dagoroot*, *Rown*, and *Huegeland*.

The chief Ports and Places of note bordering on the Sea, are *Elsinore*, strongly seated on that narrow Streight, or *Fretum*, not above a Dutch mile in breadth, commonly called by the name of the *Sound*; Over against which, on the other side, is *Elsingburgh*, a streight through which all Ships that have any trading to or from the *Baltick-Sea*, most of necessity take their course, all other Passages being either barred up with impassable Rocks, or otherwise prohibited by the Kings of *Denmark*, upon forfeiture of all their Goods. *Copenhagen*, or Haven of Merchants, placed by the Sea in the same Island of *Zealand*, being a convenient Port; This and the magnificent Castle of *Cronenburgh* near *Elsinore*, being the constant Residence of the Kings of *Denmark*.

Swedland. The next are *Sleswough*, *Elholm*, *Calmar*, *Zwideltcoppen*, *Nordcopen*, *Nycopen*, *Stockholm*, the Metropolis and chief trading Port of *Swedland*, and a place worthy observation for Merchandize; exceeding strong, both by Art and Nature, being scituate in the Marishes, like *Venice* at the Mouth of the Lake, or River *Meler*; the passage to it out of the Bay being very narrow, and yet so deep withal, that the greatest Ships of burthen may sayl up to the City; the Port within the Streight being so safe and capacious, that it is able at one time to receive 300 Sayl, which usually ride there without Anchor.

North. Bottom. Next, *Upsal*, an Arch-Bishops Sea and University, placed not far from the Bay of *Bodoue*, called also *Sinus Bodicus*, or the *North Bottom*, a long and not much frequented Sea, which from the Latitude of 60, extends it self even to the Coasts of *Lapland* and *Finmark*.

Places of note are few worthy observation, the chief *Birkara* in *West Bodden*, betwixt the Bay and a great navigable Lake: *Torneia* the best place of Trade, seated at the very bottom of the Bay in *North Bodden*: *Helsagelina* more North than that, towards the Borders of *Lapland*: *Kerlubi* in *East Bodden*, on the Bank of the Gulf, conveniently seated for a Town of Trade. The Countrey is but barely stored with Grain and Fruits, but full of great variety of Wild Beasts, whose Rich Furs yield great profit to the Inhabitants; and by reason of the commodious situation on all sides of the Bay, well stored with Fish.

Liefland. At the South-east part of this Bay is the Island *Eri*, near to the Town *Aboo*, from whence all alongst the Shores Eastward, on the South side of *Finland*, the Coast is exceeding dangerous, and for the most part innavigable, because of the innumerable multitude of Islands, Shoals, and Rocks, the greatest of which is called the *Pithing*, even as far as *Wyburg*, a Town conveniently seated at the bottom of the Bay or Gulf of *Finland*, called *Sinus Finnicus*. Over against which is *Narva*, on the North Bank of *Duina*, where it falls into the Bay of *Finland*, the only place of Trade, in-to *Moscovia* or *Russia*, through the *Baltick*.

Revel a well traded Port, scituate in the same Bay, which together with *Wyburg* and the *Narve*, are now in the possession of the King of *Swedland*.

Polonia. The next Port of note is *Riga*, a famous Empory, of great resort for Forreign Merchants; who carry hence Pitch, Wax, Hemp, Flax, and such other Commodities.

Dantzick, seated on the *Wyssel*, second of the Hanse-Towns, of so great Trade, such a noted Granary for all sorts of Corn, issued from thence to supply the want of other Countreys, that 1000 measures of Wheat (besides all other Commodities proportionable) are here daily sold.

Stetin once a poor Fisher-Town, now the Metropolis of *Pomeren*.

Pomeren. *Stralsund* a Town of much Trading, and great resort, fort, fituate on the *Baltick*, oppofite to the Ifle of *Rugen*.

Roftock next in reputation of all the Hanfe-Towns, to *Lubeck* and *Dantzick*; large, rich, and much frequented by all forts of Merchants.

Wifmar and *Lubeck*, feated on the confluence of the *Trave* and *Bilow*, near the fall thereof into the *Baltick*, a River capable of Ships of 1000 tuns, which commonly they unlade at *Travemund*, the Port Town of the City, a little lower nearer the Sea, an enfranchized Town being the principal among the Hanfe-Towns.

Jutland. On the Coafts of *Jutland*, being a Peninfula, between the *Baltick* Sea and *German* Ocean on the eaft parts, whereof there is another paffage into the *Baltick* Sea, *Belt.* called the Belt, but not fo much frequented as the *Sound*, formerly fpoken of.

The chief Towns and Places are *Flenburg*, having a Port fo deep, fo fafe, and fo commodious, that they may lade and unlade their great Ships in a manner clofe by their Houfes.

The other are *Hadersleve*, *Sternbergh*, *Slefwick*, *Wyburg*, and *Odenfee* in *Funen*, *Arhufen*, and *Sebagen*, the moft northerly point of *Jutland*.

Ham-burgh. On the Coaft of *Germany*, contiguate with the One-ao, are firft *Hamburgh*, on the Bill, where it falls in-to the *Elve*, one of the Hanfe-Towns alfo; having by report, as many great Ships as fayl upon the Ocean, which being great profit, befides the refort of Merchants from moft places. It was fometimes the Staple Town for the Cloth of *England*; on fome difcontent remo-ved from thence to *Stade*, a little nearer the Sea, on the fame River; from thence afterwards to *Holland*.

Next *Bremen*, feated on the broad and navigable Ri-ver *Wifer*, whence comes ftore of Linnen Cloth, called from a Town not far thence *Ofenbridge*.

Then *Embda*, a good Haven, and well traded Town, which yearly fends out 700 Buffes for the Herring-fifhing on the Coafts of *England*.

Ffland. Alongft the Shores, for the moft part, belonging to the States of *Holland*, lie feveral Iflands, the chief whereof are *Ameland*, *Sebelling*, *Holland*, *Fly-land*, *Texel*, *Wieringen*, *Voorn*, *Tffelmond*, *Overflacket*, *Sebonen*, *Duveland*, *Tertolen*, *North-Beverland*, *South-Beverland* and *Walcheren*.

The chief Ports and Places, are *Amfterdam*, a very fair Haven, fituate on the Gulf, called the *Tye*, and the Channel, or *Diep Anftel*, whence *Amfterdam*, built on Piles like *Venice*, and much refembling it both in Trade and other Things; a place ftored with multitude of fhipping, inhabited by Men of all Nations, and of all Religions; Grown famous, and exceeding Wealthy, fince the diverting of the Trade from *Antwerp* hither.

Horn, *Enchufen*, on the very Point of the Gulf of *Zuyder-Zee*, oppofite to *Frieyland*, *Midonblick*, *Schei-dam*, *Delf-haven*, *Rotterdam*, on a Channel named the *Rotter*, not far from which the *Lech*, one of the three main Branches of the *Rhine*, falleth into the *Maes*, a ftrong, fafe, and well-traded Port.

The firft in the Ifland *Voorn*, once Cautionary to the *Englifh*, with the Town of *Flufhing*, *Bergen up Zoom*, fo called from the River *Zoom*, on which it is fituate, about half a league from the influx of it into the *Scheld*, and not far from the Sea, which gives it a reafonable good Haven.

Antwerp fituate on the *Scheld*, feventeen leagues from the Sea, of fo great Trade in former times, that it was held to be the richeft Empory of the Chriftian World; the Commodities here Bought and Sold a-mounting to more in one month, than thofe of *Venice* in two years; the caufe whereof was, that the *Portugals* diverting the *Alexandrian* and *Venetian* Trade to *Lisbon*, kept here their Factories, and fent hither their Spices, and *Indian* Commodities, now almoft removed by the

Flanders. *Hollanders* to *Amfterdam*; *Middleburg*, *Flufhing*, the Key of the *Netherlands*; *Oftend*, *Newport*, *Dunkirck*, *Graveling*, the laft of *Flanders*.

France. On the Coaft of *France*, aboveft the *Englifh* Chan-nel, are firft *Calice* at the very entrance; *Drep* a Town of Trade efpecially for the *New-found-land*, *Newhaven*, or *Haverdegrace*, on the Mouth of the River *Sein*, be-twixt which, and St. *Maloes*, clofe by the *Hag-point*, over againft the *Ifle of Wight* in *England*, lyeth the Iflands *Alderney*, (or as the *French*, *Aurney*) *Jerfey*, *Guernfey*, belong to the Crown of *England*, and feveral other fmaller Iflands, ftored with plenty of Syder, and fine Wooll, whereof they knit ftore of Stockings and Waftcoates.

St. *Maloes*, *Morlaix*, *Rofcoaft*, *Breft*, feated on a fpacious Bay of the Weftern Ocean, the Key and Bul-wark of *Bretaign*, and the goodlieft Harbour of all *France*,

Croiffe, a little Haven at the Mouth of the *Loir*, not far below *Nants*, whence ftore of the beft and moft no-ted Brandy.

Rochel a Town feated in the inner part of a fair and capacious Bay, affured by two ftrong Forts, betwixt which there is fcarce more fpace than for a Ship to come in at once; Over againft which lyeth *Oleron*, an Ifland yielding great quantity of Salt, in fpecial fame for that the Marine Laws, which for near 500 years, have ge-nerally been received by all the States of the Chriftian World which frequent the Ocean, for regulating Sea Affairs, and deciding of Maritime Controverfies, were declared and eftablifhed here, The Ifland being then in poffeffion of the *Englifh*, from thence named the Laws of *Oleron*; So powerful were the Kings of *Eng-land* in former times to give Laws to all that traded on the Ocean.

Burdeaux feated on the *Garond*, not far from the Sea, much frequented by *Englifh* and *Dutch* for *Gafcoign* Wines; *Bayon* the laft of *France* on this part of the Ocean.

Spain. On the Coaft of *Spain*, St. *Sebaftian*, a noted and well-traded Port, at the Mouth of the River *Groinne*, beautified with a fair and capacious Haven, defended with two ftrong Caftles founded on two oppofite Rocks.

Bilbao. *Bilbao* fituate fome two leagues from the Sea, on a fair and deep Creek thereof; this (and indeed all the Coaft of *Bifcay*) ftored with fuch infinite quantities of Iron and Steel, that no Countrey yeeldeth better, or in greater plenty, called for this caufe the Armory of *Spain*; exceedingly enriched by making of Armour, and all forts of Weapons, their chief Manufacture, the *Bilbao* Blades, in fuch requeft, being brought from thence, befides great quantities of Wooll hence tranf-ported.

Coruna, by us called the *Groin*, often mentioned in our ftory of the Wars with the *Spaniard*, in Queen E-lizabeths time taken by the *Englifh*, not far from the Promontorie or Cape, called *Finis Terra*, or *Cape de Cape Fi-nis Terre*, being the moft weftern end of the then knowne World.

Bayon, not far from the Mouth of the River *Mino*, (from hence called by the *Latins* *Minium*) navigable with fmall Veffels 100 miles.

Portugal. *Porto Duero*, or *Porto Port*, at the Mouth of the Ri-ver *Duero* in the Kingdom of *Portugal*.

Lisbon, upon the great River *Tagus*, a famous City for Trafficck; the *Portugals* in all their Navigations fet-ting fail from hence, 'Tis conveniently feated for Shipping, and (excepting the Court which is here kept) inhabited chiefly by Mariners and Merchants, which of their own Countrey growth, trade in Honey, Wine, Oyl, Allom, Fruits, Salt, &c. and from *Brafil* in *America*, with great quantities of Salt and fineft Sugar, and many forts of Drugs. B 2 *Setuhal*

Setubal or St. Eves, South of Lisbon, situate on a Gulf of twenty miles in length, and three in breadth, a place of principal importance next to Lisbon.

Not far to the North-west of Cape Vincent, there are certain Islands called the Azores, in the Atlantick Ocean, subject to the Crown of Portugal, and opposite to the City of Lisbon; from which distant 250 leagues, situate between 38 and 40 degrees of North Latitude, and one of them in the first longitude, which is commonly reckoned from these Islands, as being the most western part of the World, before the discovery of America. They were so called from Azor in the Spanish Tongue, signifying a Goshawk, because multitudes were there at first found: The names are these, Tercera, St. Michaels, Fyal, Gratiosa, St. George, Pico, Corvo, Flores, and St. Maries; most of them stored with Flesh, Fish, and a sort of Wine not very good, nor durable. But the chief Commodity they lend out, is Wood for the use of Dyers.

St. Lucar the Port Town of Sevil, at the Mouth of the River Betis, or Guadalquiver, where the West-India Ships many times ride.

Cadiz or Gades, situate on a large Bay, and serving as a Road for the Indian Fleet; by reason whereof, and the great resort of Forreign Merchants, it is much enriched: 'Tis the chief Port and Magazine of Spain, taken notwithstanding in one day by the English, under the command of the Lord Effingham, the Earl of Essex, and Sir Walter Raleigh; the Town, Ships, and all becoming a prey to the English.

Near to this place is that so celebrated Streight, called Fretum Herculeum, or Gaditanum, now the Streights of Gibralter, from a place so called on the brink hereof, being in length fifteen miles, and in breadth seven, where is its narrowest, being the Inlet or Passage from the Atlantick-Ocean, into the Mediterranean Sea.

The Mediterranean Sea, within which the places most observable are, Malaga, a strong place, and an Attorney for the King of Spain; exceeding great in Traffick, and of much resort, especially for Wines, Raisins, Almonds, &c.

Almeria, Carthagena, situate in a demy Island in the very jaws of the Mediterranean, having a good and capacious Haven.

Alicant a noted Port and much used, whence our true Alicant Wines, made of the Juyce of Mulberries.

Valentia, a fair, pleasant, and well-traded City. Tarragona, Barcelona.

The Goods and Merchandize on this side of Spain being generally Corn, Wine, Oyl; all sorts of Fruit, Salt, Corral; several sorts of Drugs and Stones, &c.

Over against Valentia lie several Islands, the biggest whereof are Majorca, the chief Town whereof is so named of the Island, yeelding sufficient quantity of Corn, Oyl, Wine, and Fruits.

Minorca having three fair Harbours, Maon, Ternussus, and Minorca; a fruitful Island, breeding great Heards of Cattel, and Mules of the largest size in Spain.

Yvica, the Inhabitants whereof make great store of Salt, wherewith they furnish, in part, not onely Spain, but Italy also.

Next, on the Coast of France, are Narbon, Arles, Marselles, and Thelous.

The Commodities sent from the Coasts of France, on the Mediterranean, are Corn, Wine, Oyl, Salt, Woad, Allomnet, or Grain d' Escarlate, Saffron, Rasins, Figs, Olives, Almonds, Prunes, Capers, &c.

Villa Franca, and Savona, belonging to the State of Genoa.

Genoa the principal Empory, next Venice, of all Italy, having a safe and commodious Haven: The Country Commodities are (besides their Fruits which here are excellent) Oyl, Paper, Wines, and such abundance of Silk, that it is the opinion there are 18000 persons in that only City imployed in ordering and working thereof.

Livorno, or Legorn, seated on the influx of the River Arno, so well fortified, that it is thought to be one of the strongest Cities in Christendom. To the south-West whereof, in the Ligurian Sea, lye the Islands Corsica and Sardinia; the first subject to the State of Genoa, the latter to the Kingdom of Spain; abounding in pleasant Wines, Oyl, Olive, Mastick, Sulphur, Allom, Wax, and Honey.

Elba an Island between Corsica and the Main, producing Load-stones of a gray colour, but none of the best.

Civita Vechia the onely useful Haven that belongeth to Rome.

Tarracina, or the Bay of Mula, and Port Ostia at the Mouth of Tyber, scarce making up one good Haven.

Naples the Metropolis of the Kingdom; a beautiful City seated on the Sea-shoar, and fortified with four strong Castles. This, and indeed all the Parts of Italy generally, abounding with all sorts of Silk, Cloth of Gold and Silver, made by the People without fraud, because of a strict prohibition for the Adulteration of the Threads; Topestry, Skins bravely gilded; Earthen Vessels most curiously wrought with Images and Coats of Arms; Oyl Olive of the best sort; Saffron, Alchermes, Allom, Sulphur, Vitriol, Alabaster, Rice, Marble, Wines, and Fruits of all sorts.

Regium, or Rezo, on the Sea-shoar, opposite to Messina in Sicilia, which is suppose'd to have been broken off from the Coast of Italy, a place heretofore very well traded, but since fired by the Turks, left almost desolate.

Sicilia an Island separated from the Main Land of Italy, by the Strait or Fare of Messina, where the Passage is so narrow, that it exceeds not in breadth a mile and a half, and sound, by diligent sounding, not above eight fathom deep; full of dangerous Rocks and Whirlepools: as namely Caribdis, a Gulf on Sicilia side, violently attracting all Vessels coming nigh to it, and devours them; opposite whereunto stands that dangerous Rock Scylla, at the foot of which many little Rocks shoot out, these two being the occasion of many fabulous Stories. In the other parts where the Sea opens, it is 300 miles over, supposed to have been once a Peninsula, afterwards separated from Italy by the fury of the Waves, or violence of some Earth-quakes, which are there frequent. The Island is so plentifully stored with Corn, that it heretofore obtained the name of the Granary or storehouse of Rome, and doth still furnish not onely many parts of Italy, but Spain, Barbary, Malta, and the adjacent Isles; the other Commodities are much like those of Italy, in great plenty.

In this Countrey is the Hill Hybla, so famous for Bees and Honey; the Mountains Ætna, now Mongibel, which continually sends forth smoak and flames of fire, to the astonishment of beholders.

The chief places are Syracusa, or Saragosa, once the Metropolis of the Island; very strong both by Sea and Land, with a beautiful and commodious Port, of greatest Trade, next to Carthage in antient times, now both destroyed.

Note, Augusta, Gregento, Palermo, Trapani, and Messina, a Port and City of great strength and beauty; peopled by the wealthiest sort of Merchants and Gentlemen; having a strong and high Citadel, well garrisoned, and a Lanthorn with lights kept burning for direction of Mariners.

Sixty miles to the southward of Sicilia, towards the African Shores, lyeth Malta, an Island famous for the shipwrack of Paul; defended by the Knights of Jerusalem removed hither; it is wholly situate on a Rock,

Isw ing

having not above three foot depth of Earth, and consequently of no great fertility, the want of which is supplyed by the plenty of *Sicilia*.

To return therefore to the Coast of *Italy*, by the Capes *Spartiventi*, *Colonne*, and *St. Maries*, next unto which is *Gallipoli*, noted for the excellent Oyl coming from thence.

The Gulf of Venice. Not far from whence is *Cape Otranto*, the entrance into the *Adriatick* Sea, or Gulf of *Venice*, and the first Town of note therein is *Brundisi*, or *Brundusium*, once glorying in the most capacious Haven of the World. Whence *Pompey* and *Cæsar* took shipping with their Fleets, the one to fly, the other to pursue; at this time a mean Town, the Haven being so choked up, that a Galley can hardly enter.

Next *here*, *Ortona*, *Ancona*, having a fair Haven, not so capacious, as exceeding pleasant and beautiful.

Venice. *Pesaro*, *Ravenna*, and in the top of the *Adriatick*, *Venice*, standing upon seventy two little Isles, but joyned together by many Bridges, which are said to be 4000 at the least, besides 10000 Boats for passage from Isle to Isle; a strong, beautiful, and famous City, once the most illustrious Emporie of the World, but much decayed in its Trade since the Passage by Sea was found to *Persia* and *India* by *Cape bon Esperanza*.

Dalmatia. On the *Histrian* and *Dalmatian* side of the *Adriatick*, are these places and Sea-ports observable, *Trieste*, or *Tergestum*, whence the Bay adjoyning is so called; *Zara* in *Dalmatia*, enjoying a late and large Port belonging to the State of *Venice*, *Sibenico*, *Spalatro*, *Narenta*, *Cattaro*, at the bottom of the Gulf so called.

Grecia. On the Coast of *Greece*, over against *Otranto* in *Italy*, lieth *Valona*, a Port Town, fortified with a strong Castle.

Farther into the *Ionian* Sea lie several Islands, first *Corfu*, sufficiently wasted, but of Wine especially,

Cephalonia. Zant. *Cephalonia* and *Zant*, Islands abounding in Wine, but especially in Currants, which is the greatest trade of these Islands.

Larta, on the Grecian Shore, in a Gulf, so called, near the antient *Ambracia*, the Regal Seat of King *Pyrrhus*; near unto which is the Isle of St. *Maure*, inhabited chiefly by Jews; a little lower than which is the Gulf of *Corinth*, called *Lepanto*, from two Castles built on each side the entrance thereof, called *Castello de Lepanto*, made famous by the memorable Sea fight of the Turks and Christians.

Morea. *Peloponesus*, now *Morea*, a Peninsula joyned to the Continent by a little neck of Land or Isthmus, at *Corinth*, six miles over in breadth; the pleasantest Country of all *Greece*, abounding with all things necessary for life, now in possession of the Turks; and though no place hath suffered more ruine than this, yet is it still the most populous of all *Greece*. The chief places are, *Modon*, or *Methone*, seated on the most southern part of the Peninsula; a strong, safe, and convenient Harbour.

Coron the chief Town on the Bay of *Messina*.

Malvasia, anciently *Epidaurus*, noted for the abundance of delicate Wines, called Malvesy, or Malmsey, first hence into all ports, *Napolia*, now *Napoli*, giving name to the Bay so called.

Candia. To the South-east hereof lyeth that famous Island of *Crete*, now *Candia*, from the chief Town so called, abounding heretofore much more than now, in Muscadel Wines, Oyl, Sugar, Gums, Honey, and Fruits. The People formerly good Seafaring Men, subject to the *Venetians*, till much rent from them by the Turks, especially of late, by the strong Town of *Candia*, situate on the North Coast.

The next places of note, whereof are *Retimo* and *Canea*, commodious by its Haven, called *Porto del Suda*.

In and about the *Ægean* Sea, lie many Islands, the most of note are, *Samothrace*, now *Samandrachi*; *Thassus*, or *Tasso*; *Imbros*, now *Lembro*, *Lemnos*; the Merchandise whereof is that Mineral Earth called *Terra Lemnia*, and *Sigillata*, from the Seal or Character imprinted on it.

Archipelago. *Euboea*, now *Negropont*, full of Harbours and capacious Bays.

Salamina, *Ægina*, or *Engia*, the *Cyclades* so called, because placed in a Circle; called also the Arches, the chief whereof are, *Delos*, *Tenos*, *Andros*, *Naxos*, *Gyaros*, *Paros*, *Styros*, *Milos*, *Scriphos*, *Cea*, and several others, in number fifty three. The *Sporades*, because scattered up and down the Archipelago, in number twelve. Lastly, *Cythera*, now *Cerigo*.

On the other side next *Asia* lie *Tenedos*, *Lesbos* or *Mitilene*, *Chios*, *Samos*, *Coos*, *Icaria*, *Patmos* now *Palmosa*, *Claros* now *Calamo*, *Carpathos*, *Rhodes*, memorable in the hard and long Siege of the Turks before it was taken.

In this mentioned Sea, on the Continent Shores of *Greece* and *Thrace*, are observable first *Athens*, now called *Setines*; so famously heretofore memorable, now an ordinary Borough.

Next *Thessalonica*, now *Salonichi*, at the bottom of a great Bay so called, a beautiful wealthy City, inhabited by rich Merchants of most Nations and Religions, who drive here a great Trade.

Abdera, *Ænos*, *Lysimachia*, and *Philippopoli*, on the River *Hebrus*.

Selimbria, *Sestos*, a Peninsula in the *Thracian Chersonese*, having a strong Castle; opposite to which before is another on the *Asian* Shore, called *Abidos*, both of them having the name of the *Dardanells*, the Key and Blockhouses of *Constantinople*, commanding the Passage so strongly, that none may go out or in without their licence.

Constantinople. But the chief glory of this Country and of all *Europe*, is *Constantinople*, seated in so commodious a place for Empire, that it overlooks both *Europe* and *Asia*; commands not onely the *Propontis* and *Bosphorus*, but the *Euxine* Sea; first called *Byzantium*, now since the possession of it by the Turks, *Stamboldi*, having a most curious Haven or Port so conveniently profound, that Ships of greatest burthen may lie at the sides thereof for recone and discharge of their Lading; so conveniently seated, that there is no Wind whatsoever but brings in some Shipping; which affords a vast trade of Merchandize from all parts, and of all sorts.

Euxine Sea. Beyond this is the *Propontis* and the *Euxine*, or Black Sea, or *Pontus*, now *Maggiore*, a very dangerous Sea, full of Rocks and Sands, guarded at the entrance by the *Bosphorus* with two strong Castles, called the Black Towers.

Smyrna. But to return again to the *Ægean* Sea, by the Coasts of *Asia*, the places most worthy of notice for Trade are here, the chief *Smyrna*, a fair and antient City, on a large Bay to named, much traded and frequented at this day, especially for Chamlets, Grograms, and such like Commodities, where the English have a Consul resident.

Ephesus, *Halicarnassus*, *Matari*, *Antioch*, of no great Trade.

And at the end of the *Mediterranean* Sea, *Alexandretta* or *Scanderoon*, pretty commodious for Trade, being the nearest Haven to *Aleppo*, heretofore the choice Staple for all the Eastern Commodities brought to *Euphrates*, before the *Portugals* discovery of the Southern Passage to *India* and *Persia*.

Tyre a City in antient time of great Trade and Wealth, seated on a Rocky Island, the People whereof were supposed to be the first that invented shipping, now nothing but a heap of rubbish.

Tripolis,

Cyprus. *Tripolis*, over against which is the Island *Cyprus*, in the *Syrian* and *Cilician* Sea, abounding in Wine, Oyl, Corn, Sugar, Cotton, Hony, Wooll, Turpentine, Allom, Verdegreece, Salt, Grograms, and other Commodities.

Egypt. On the Coasts of *Egypt* and *Barbary*, are first *Alexandria*, situate westward of *Delta*, over against the Isle *Pharos*, at the Mouth of the River *Nile*; exceeding strong, inhabited by men of divers Nations, as Moors, Jews, Turks, Greeks, and almost all other Nations, for the gain they reap by trafficking in Corn, Rice, Estridge-Feathers, Gums, Drugs, Spices, Cotton-Cloth, and other rich Commodities.

Tripoli in Barbary, an usual recess for Pyrates that infest those Seas.

Next *Tunis*, whose Commodities are chiefly Oyl, some Corn, Figs, Dates, Almonds, and other Fruits.

Argiers, situate near the Sea, in the form of a Triangle, with an Haven to it, a City not so large, as strong; and not so strong as famous, for being the Receptacle of the *Turkish* Pyrats, who domineer so infinitely over the *Mediterranean*, to the great dammage of Merchants that frequent those Seas.

Tetuan, the last Town within the Straits on the *African* Shore.

Tanger. On the *African* shore without the Straits, lyeth *Tanger*, near *Cape Spartel*, a Town very antient, thirty miles distant from the Straits Mouth; belonging heretofore to the *Portugals*, now to the Crown of *England*, where there is lately built a large and convenient Mole for the reception of Ships; and a strong Garrison for defence of the place, and against the incursion of the Moors.

Salley, a Town much traded formerly by Merchants of *England*, *Flanders*, *Genoa*, and *Venice*, of late made a nest of Pyrats, as dangerous to those that sayl in the Ocean, as *Argier* to those in the *Mediterranean*.

South-westwards from hence are the *Canaries*, or fortunate Islands, in number seven, so called from *Canaria*, the principal thereof: The names are these, *Canaria*, *Palma*, *Gomera*, *Ferro*, *Lancerota*, *Tenariff*, and *Forteventura*; called Fortunate, from their fruitfulness and other excellencies, plentiful in Wood and other Commodities, but especially in those rich Wines we call Canaries; a sort of Wine, if not sophisticated, more pleasing to the Pallat, and a better Remedy for the natural weakness of the Stomach, (if moderately taken) than any other Wine whatsoever; brought hither in such abundance to supply our luxury, that much more than three thousand Tuns hereof are brought yearly into *England* only.

Canaries

Next *Madera*, the greatest Island in the *Atlantick*-Sea, over against *Cape Cantin* in *Morocco*, wonderfully fruitful, abounding in Madder, Sugar, Fruits, Wines, &c.

Not far from thence is the Isle *Porto Sancto*, very fruitful also, but much annoyed by the innumerable multitude of Conies that breed there.

About *Cape Blanco* generally the Sea-Ports and Places, even to the farthermost parts of *Guinea*, yea even of all *Africa*, were belonging to the *Portugals*, who fortified and placed Colonies in each as their trading increased; as *Porto de Dio*, *Del Rosato*, *Arguin*, *Rio de Portugaez*, or *Senega*, betwixt which and the River *Gambo*, is the great westward Cape of *Africa*, called *Cape Verde*, then *Rio de Sancto Domingo*, and *Rio Grand*, &c.

Now several of them are much used and frequented by Dutch, English, and French. The Commodities are chiefly Gold, Ivory, and a sort of Pepper which we call Guinea Pepper, of double efficacy to the *Indian*.

The Isles of Cape Verde. To the Westwards of *Cape Verde*, lie the Islands, so called, being ten in number, St. *Antonio*, St. *Vincent*, *Bona-Vista*, St. *Lucya*, the Isle of *Salt*, *Del Fogo*, St. *Nicholas*, *Mayo*, St. *Jago*, and *Brava*.

Here the continuance of this Discourse should have been broken off to have took in the Northern Tract of *America*, as far as the Equinoctial, so to have compleated this Hydrographical Description, according to the first division of the Ocean; but I thought it more convenient to go forward through the remaining part of the *Ethiopian* and all the *Indian* Seas, alongst the Shores of *Africa* and *Asia*, and having briefly spoken thereof, to comprehend all the West *India* or *America* in one Treatise.

To proceed then on the Coast of *Africa*: from the River *Gambo*, to the *Cape of Good Hope*, are the Coasts of *Malegette*, the *Grain* Coast, *Cape de Palmas*, *Quaqua* Coast, the Gold Coast, *Cape de tres Puntas*, the Coast of *Benin*, called also the *Byte* or *Gulf* of St. *Thomas*, *Cape Formosa*; all this whole Countrey abounding in Corn, Rice, Millet, excellent Fruits; also in Gold both in Sand and Ingots, Ivory, Wax, Hides, Cotton, Ambergreece, Brasil-Wood, Pearls; which they truck for Cloth, Woollen and Linnen, Red-Caps, Frize-Mantles, Guns, Swords, Daggers, Belts, Knives, Copper-Bars, Hammers, Ax-heads, Salt, Pins, Kettles, Basons, Looking-Glasses, Beads, Tinn-Rings, and certain Shells called Cories, which passeth there instead of Money. They drive a great trade for these said Commodities with their own people, whom they sell for Slaves, the Kings selling their Subjects, Parents their Children, and indeed all whom they can take or surprize, which are sent generally to the *West-India* Plantations. Guiny.

To the southward hereof the Ports are divers, but little frequented by the English. The *Portugals* conquering and possessing several places from the weak Native Inhabitants, all along those Shores. Much thereof being since gotten by the Dutch, some by us and others; all which do generally abound with the usual Merchandize of the other western parts of *Africk*.

Here also must not be omitted the mention of such Isles as lie in this part of the *Ethiopian* Sea, namely, St. *Thoma*, just under the Equator, inhabited now by the Dutch, *Ferdinand de Po*, *Princes Island*, *Anabon*, St. *Helena*, the usual place of Watering in the return of *East-India* Voyages, being in possession of the English *East-India* Company; *Ascention*, a barren Island, whereunto sometimes Ships go a turtling

Cape de bon Esperanza, or the Cape of Good Hope, was first discovered by *Vasques de Gama* a *Portugal*, Anno 1597. by which Discovery, monopolizing to themselves the wealthy Trade of *India* for a great while, till by one means or another communicated to others. The Cape consisteth of three Points or Head-lands, whereof that which is nearest is called as before; the middlemost, *Cabofalso*, because mistaken for the other by some of the *Portugals* in their return homwards; the other the *Cape of Niedles*, or *Cape das Agulhas*, by reason of the sharp Points it shoots out into the Sea. On the top of this Cape is a large and pleasant plain, called the Table of the Cape, yeelding a large prospect over the Sea on all sides. Cape of Good Hope.

Beyond which, the first Port of observable note, is *Sofala*, on a little Island near the great River *Cuama*, next *Mozambique*, conveniently seated on a large and capacious Haven; strongly fortified, in the hands of the *Portugals*, who in their going to the *Indies*, and returning back, used to call here, and to fit themselves with all things necessary to pursue their Voyages: A Town of so great Trade and Wealth, that the Captain of the Castle, in the time of his Government, being but three years, is said to have laid up 300000 Duckets

for

for his lawful gains, out of the Gold, &c. there, and coming from Sofala.

Over against this Port eastward, lyeth the great Island of Madagascar, or St. Lawrence, being the greatest yet known in the World; plentiful in all things for the life of man, particularly of Mill, Rice, Sugar, Honey, Wax, Cotton-Wooll, Coco-Nuts, Dates, Goats, Deer, Oxen, Sheep, Fruits, Ginger, Cloves, Sanders, Saffron, Amber, Gold, Silver, Ivory, and Ebony; which they exchange for Toys and small Trifles. The Inhabitants inhospitable and treacherous. Harbours it hath many, and often frequented by Portugals, Dutch, and English.

Up higher towards the Arabick-Gulf, are Melinde, Mombaza, Quiloa, Mogadoxo, &c.

At the most eastern port of Africk, called Cape Gardofu, lyeth the Island Zocotora, abounding in Cinnabar, Dragons-Blood, and Alloes, hence called Aloes Socotrina.

Here is the entrance into the Arabian-Gulf, or Red-Sea, rightly so called from bordering on the Land of Edom. The chief Ports whereof are Zeila, Maczua, on the Coast of Ethiopia. And at the very top thereof Suez, or Arsinoe, the station of the Turkish Gallies that command the Gulf, they being first framed at Cairo, then taken in pieces, brought hither, and here rebuilt and joyned together.

Eziongeber, the Haven of Solomon's Ships, that fetcht his Gold from Ophir.

Others in this Gulf, on the Coast of Arabia, are few, or no places worthy of mention, as far as Babel Mandel, where it openeth into the Southern Ocean; most part of the Persian and Indian Merchandize coming formerly this way, and so transported by Land to Cairo, then to Alexandria, but now little or nothing used.

A little without the Gulf standeth Aden, a gallant Haven, well traded, and seldom without store of Shipping, carrying from thence Gums, Drogs, and other Merchandize.

Next Ormus, the Lock and Key of the Southern Ocean, on the Point or Promontory, at the entering into the Persian-Gulf, or Gulf de Eluatiffe, a turbulent and unruly Sea, the Southern Ocean breaking in at one end, and the River Euphrates at the other, the combined combating and clashing of which two, makes it so unquiet.

Places and Ports of note on the Arabian Shore, are Musquat, Sabes, Balsara, Rheginae.

But none so famous as the City of Ormus, on the Persian Shore; not so memorable for the greatness, as the wealth and conveniency of the situation thereof; built in an Island, so called, a famous Empory for Persian and Indian Commodities; being hence transported and conveyed to Badgar, or Babylon, Aleppo, and Tripoli, not yet wholly decayed; besides plenty of other Merchandize, here are found the best and fairest Oriental Pearls, which are caught in this Gulf between Bafora and Ormus.

The first Port on the Coast of India, is accounted Diu, looking towards Persia; but on the East side thereof, near the Mouth of the River Indus, a Town of great Trade, possessed by the Portugals.

Tutta on the Banks of Indus, of no less trade to the Portugals, who here receive such Indian Commodities as come down the Water from Labore, returning Pepper in exchange, which they bring up the River from their other Factories.

Madabar the chief Town of Guzarat, affirmed to be near as big as London; seldom without Merchants of all parts.

Cambaia, 3 miles from Indus, and as many in compass, so populous, that it is accounted the Cair of the Indies; exceeding fruitful, abounding in Rice, Wheat, Sugar;

all sorts of Spices and Fruits, Silk, Cottons; and in the Mountains thereabouts they find Diamonds, Calcedonies, and a kind of Onyx, called Corneline, corruptly Cornelian.

Swaley, in a large Bay so called, the Haven Town for Surat, about ten miles from the Road, from whence the River is Navigable only by Boats and Shallops; made of late years a Factory for the English Merchants, who have here their President, and a Magnificent House for their Reception, and Staple of their Commodities, which are chiefly Spices, Calicoes, Indico, Sarcinets, Pantadoes, &c.

Bombay is a great Factory for the English East-India Company.

Goa a Sea-Town, situate in a little, but most pleasant Island, called Tisuarinam, fifteen miles in compass, opposite to the Out-let of the River Mandae, a noted Empory, and one of the chief Keys that unlock the Indies; inhabited, besides the Portugals, by Indians, Moors, Jews, Armenians, Guzarats, Banians, Brasilians, and many others, who for the cause of Trade and Gain, dwell here, without molestation for their Religion.

Cunnar Batticale, on the Coast of Malabar, first Conmus; well built and beautified, with a very fair Haven, belonging to the Portugals, and well traded by Merchants.

Then Calicut, the chief of these parts, three miles in length upon the Sea, of exceeding Trade, especially in fine Calicoes, thence so called, Ginger, Cinamon, Pepper, and Cassia.

Cochin, a Sea-Town likewise, of little less Trade than Calicut.

To the southward hereof is Cape Comorin, or Cormandel; and a little from thence the Island Zeylan, large, and almost round, affirmed to be plentiful in Cinnamon, Ginger, Gold of the best sort, Silver, and all sorts of Mettals, Precious-Stones, and store of the largest Elephants; the chief Towns are Triquelimole and Batticala, Jaffanapatan, Colmuch now Colombo, having a fair Haven, the Royal Seat of the Kings, whence many Ships laden with Cinnamon, Gems, Elephants, and other Commodities go yearly to other places.

Within the Gulf of Bengala, in the Kingdom of Golkonda, are Nagapatan, Madras, St. Georges Fort, Masulapatan, Orissa, Bellasor, Angeli, &c. From all which they chiefly send plenty of Rice, Cotton-Cloth, a fine Stuff like Silk, made of a Grass, called these Teron; Long-Pepper, Ginger, Mirabolans, and other Merchandize.

Ougely and Bengala, giving name to the great Bay, situate on a Branch of the River Ganges, a place endowed with plenty of all things fit for life, rich in Merchandize; especially Rice, Gold, Precious Stones, Pearls, a curious sort of painted Cotten Cloth thence sent to all parts of the World.

Aracan and Pegu, the glory of these parts, great, strong, and Beautiful, Rich in Gold, Gems of divers sorts, Red-Wax, &c.

Lugor, on the Sea-side, near that little Isimos that joyneth the Chersonese to the main Land.

Martaban, Sornan, Queda, renowned for the best Pepper, and in most plenty, for that cause much frequented by Merchants.

And in the Kingdom of Siam, in the narrow Strait between the Isle of Sumatra and the Peninsula, called the Golden Chersonese, stands Malaca, for Spices, Unguents, Gold, Silver, Pearls, and Precious-Stones, the most noted Empory of the East, once possest and strongly fortified by the Portugals, but taken from them by the King of Achem.

Next Jor, at the very Point or Promontory, where the English and Hollanders have their Factories.

Siam, at the bottom of a great Bay, a goodly City, and conveniently seated on the River *Menam*, for Trade and Merchandize, which is Precious-Stones, much Spices, &c.

Cochin-china. *Champa* and *Cochinchina* in *Camboya*, having store of Gold, and Lignum Aloes, valued at it's weight in Silver; Silk in abundance, Purslain Earth for the making Cups, Dishes, and other Utensils, Salt-Peter, &c.

To the northward whereof lyeth the Isle of *Aynao*, a place of the greatest note for the Pearl-fishing.

China. Hereabout beginneth the Kingdom of *China*, which as it is the largest, richest, and best inhabited throughout the whole World, would require a Treatise correspondent; but because they are a People forbidding Foreigners to trade amongst them, unless in some few places, the knowledge of others coming onely by particular report, I shall only give a touch at two or three places which are most considerable for Trade; as

First, *Nanquin*, of incredible greatness, situate in a great Gulf, so called, nine leagues from the Sea, on the great River *Kiang*, wherein, by report, ride for the most part no less than 10000 of the Kings Ships, besides such as belong to private Merchants.

Canton, on the Navigable River *Macao*, where the *Portugals* had once a great Factory.

Paquien, not far from the Sea, well-traded, and conveniently seated for conveyance of Merchandize throughout the whole Kingdom.

Sicunbuy a Town frequented by much Shipping, not above twenty four hours sail from *Japon*, the Trade whereof is chiefly Cottons.

Numberous are the Ports of *China* besides these, which for the cause aforesaid are omitted: The general Trade whereof consists chiefly in Gold, Silver, Copper, China-Silks in abundance; fine Porcelain, Rhubarb, Musk, Civet, Amber, Camphire, Spices, Pearl, much China-Wood, and almost all sort of Merchandize.

The Islands scattered up and down the *Indian* Seas, are very many, and rich in Merchandize, &c.

Japan the most northern Island of all, having several fair Ports, *Meaco* being the chief, or this time the Empory and Staple of *China*, whither they bring their Commodities for foreign Trade.

Philippines. The *Philippines*, so called, in honour of *Philip* the Second, King of *Spain*, in whose time discovered, many whereof have been, or now are, under that Crown.

Mauilan Manaw, unfortunately remarkable for the death of *Magelan*, there slain in a Battle with the Natives.

Molucca. *Lequio Major*, *Lequio Minor*, *Formosa*, *Reta Major*, the Isles of *Bandan*, *Moluccoes*, *Ternate*, *Tidor*, *Mather*, *Rachoan*, *Machian*, *Batene*, *Celebes*, *Gilolo*, *Maccassar*, and *Amboina*, where that inhumane Butchery was by the Hollanders committed upon the English, *Anno* 1618. Upon all which respectively, not onely the Merchants of *China* and *India*, but the Portugals, Spanish, Dutch, and English, have continual recourse by Shipping, bringing from thence Gold, Silver, and other Metals; Gems, Pearls, Nutmegs, Mace, Cloves, Cinnamon, Ginger, Aloes, Sugar-Canes, Pepper, Drugs, Sanders white, red, and yellow, &c.

Borneo. *Borneo*, an Island of more note, and greater than any other spoken of in the Indian Sea, just under the Equator; the greatest riches whereof are, Camphire, Agarick, and Diamonds.

Plates of note therein see, *Baraca*, *Sambar*, *Succadana*, *Benjamassin*, &c.

Sumatra. *Sumatra*, under the Line also, whence to the Coast of *Malacca* the Strait is very narrow, nor above a Musquet-shot in breadth; it affords great plenty of Wax, Silk, Cottons, Ginger, Pepper, Camphire, Agarick, and Cassia; rich in Mines, not onely of Tin, Iron,

Sulphur, and other Minerals, but of Gold such plenty, that 'tis credibly believed this was the Ophir of *Solomon*. The Inhabitants are either good Artificers, cunning Merchants, or expert Mariners. The chief Sea-Towns, *Achim*, the Royal Seat; *Pedir*, *Camber*, *Menancabo*, and *Passaman*.

Java Major, rich in Corn, Metals, Gems, Silks in abundance; Pepper, Ginger, Cinnamon, and store of their Spices. The chief Towns, *Palucbun*, *Sacubaya*, *Tuban*, *Dama*, *Charabon*, *Batavia*, and *Bantam* near the Straits of *Sunda*, which separate *Sumatra* from this Island.

This place, amongst many others, being the principal Factory of the *English* in all that part of the *Indies*.

The King of *Bantam* having great correspondence with, and great affection for, his Majesty of *England*; whereby 'tis hoped our Factory will be better feeled, and our Traffick encreased in those parts, to the great advantage and profit of our English Merchants.

Japan-Minor, the South Coast whereof is not fully discovered, and the Places and Commodities onely by conjecture, so also are many other Islands and Places thereabouts, as *Nova Hollandia*, *Nova Guinea*, *Islas de Ladrones*, &c.

America. In the *Indian* Sea, and *Mare Pacificum*, which with the Coasts of *America*, remains onely to be spoken of. That great Sea or Ocean, was first so named by *Magellan*, who passing through those troublesome and tempestuous Seas, that bear his name, found such a change upon his coming into this main Ocean, that he gave it the name of *Mar del Zur*, from the calm and peaceable temper thereof.

California is the most Western part of *America* which is washed by this Sea, once supposed to be a part of the Continent, but since discovered to be a large Island separated from the Main by a narrow Sea called *Mar Vermeglio*, by some, the Gulf of *California*.

Towns of trading here are few or none, at least onely to us known; the Capes only observable, once coasted by Sir *Francis Drake*, as *Cape Blanco* and *Mendocino* in the North, and St. *Lucas* on the South, remarkable for the great Prize taken there from the Spaniards by Capt. *Cavendish*, in his Circum-navigation of the World.

New Spain. On the South-east hereof are the Navigable Rivers of St. *Sebastians*, *Rio de Spirito Sancto*, *Capa Corrientes*; the Town of *Natividad*, pillaged and burnt by Capt. *Cavendish*; St. *Jago*, a little South of *Natividad*, the Shores whereof are said to be full of Pearls.

Acapulco the best Haven on the South Sea, in a safe and capacious Bay, with convenient Stations and Docks for shipping.

Aguatulco a noted Port, and rich, much used in the Spanish Voyages from *Mexico* southward, plundered by Sir *Francis Drake* and Mr. *Cavendish*.

Tacoante peque, *Guatimala*, and St. *Jago*, *Salvador*, St. *Michaels*, *Carlos*, *Philippina*, St. *Foy*, where the Spaniards melt and cast their Gold into Ingots.

These, and indeed all the Western Shores of *America*, subject to the Spaniards, they being very cautious and jealous of any other Countrey to trade there; many of these Countreys, especially the Valleys, exceeding fruitful in Fruits and other necessaries for life, the mountainous parts being barren, but plentifully supplyed with never-perishing Mines of Silver and Gold; the other Merchandize being Cottens, Sugars, Indigo, Cochineel, Liquid-Amber, Mastick, Tobacco, Sulphur, Sarsaperilla, several sorts of Gums, and other Apothecaries Drugs.

A little beyond Cape *Saulla Maria*, in *Veragua*, lyeth *Panama*, over against *Porto bell*, being the narrowest part of that long and narrow Isthmus, or Strait of Land that parteth the two Peninsula's of *America*, Mexi-

cana

tana and *Peruana*, called the Straits of *Darien*, from a Town and River of the same name; in some places not twelve miles from Sea to Sea, in many not above seventeen; a small Ligament for so great a Body, observable by that notable but successless attempt of *John Oxenham*, an adventurous Englishman, one of Sir *Francis Drake's* Followers, who arriving with seventy of his Companions in a small Bark, a little above *Nombre de Dios*, the chief Town of the Isthmus, or *Mar del Nort* side, drew his Ship on Land, covered it with boughs, and guided by some Negroes, marched over-land with his Company, till he came to a River; there cut down Wood, made him a Pinnace, entred the South-Sea, went to the Isle of *Pearls*, took from the Spaniards 60000 pound weight of Gold, and 100000 pound weight of Silver, returned to Land; but through the mutiny of some of his own Company, was intercepted, and never returned to his Ship or Countrey. The Recorded by the Spanish Writers with great admiration.

The Towns and Places on the Shores of *Peru*, are first *Bonaventura*, on a Bay so called; *Cape de Francisco*.

Peru. *Puerto Viego*, not far from the Sea, the first Town of these Parts possessed by the Spaniard, from whence the Trade is driven betwixt *Panama* and *Peru*, where are found very rich Emeralds.

Guyaquil, in a deep Bay, a noted and much frequented Emporie of the Spaniards.

Payta, a small Town, but hath the safest and most frequented Harbour in all this Countrey: born by Captain *Cavendish*.

Lima, by the Spaniard called *Ciudad de los Reyes*, the most fruitful of all *Peru*, in the Latitude of 12 deg. and a half; a Town of greater wealth than bigness, the Riches of *Peru* passing yearly through it: sacked by Sir *Francis Drake*.

Porto Quemado, *Castro Vereyna*, whence cometh that Tobacco called Right *Vereyna*.

Ariquipa, where the Silver of *Plata* and *Fotosi* are ship't for *Panama*.

Next in *Chili* are *Copyapo*, *Serena*, *Port and Paraiso*; out of which the English under *Drake* took a Ship, and therein 2500 Pesoes of the purest Gold of *Baldavia*.

Conception, *Auraca*, *Imperial*, *Baldavia*, *Osorno*, *Castro*, the most southern Town of all *Peru*.

To this Southern part of *Chili*, there is great expectation of an English Traffick with the Inhabitants, by reason of a disgust taken by the Natives against the Spaniards for their cruelty and infidelity.

And, in order thereunto, his Majesty and Royal Highness the Duke of *York*, and several others of the Nobility, designed a farther discovery of those parts for procuring a Trade and Commerce with the People of that Countrey; and in the year 1669, there were two Ships sent upon the same Discovery, the one called the *Sweepstakes*, under the conduct of that ingenious and venturous Commander, Capt. *John Norborough*; and the other the *Batchelor*, Capt. *Humphrey Flemming* Commander. Both which Ships proceeded on their Voyage til they came near the Straits of *Magellan*, not far from *Rio St. Julian*, which was the appointed place of wintering, until an opportunity presented to pass the said Straits: near which place they lost one another: whereupon the *Batchelor* returned home, with a strong apprehension that his Consort was lost: But on the contrary, the *Sweepstakes* very honourably proceeded on her Voyage, and passed through the Straits into *Mar-del-zur*, alias, *Mare Pacificum*, and sayled alongst the Coast of *Chili*, unto a place called *Baldavia*, in the latitude of 39 deg. 30 min. or thereabouts, under the power and jurisdiction of the *Spaniards*, who have the command thereof as far as the reach of their Guns; who at first pretended a friendship with our Men, but at last betrayed and detained four of them, which Captain *Norborough* very diligently endeavoured to release, but proving ineffectual, was constrained to leave them behind, and so returned back through the Straits, and in *June* 1671 came home, to the great satisfaction of the whole Court, giving great hopes of procuring a Trade in those Parts, that may possibly prove very advantagious to the whole Kingdom, by reason of the abundance of Gold and Silver in that Country.

From *Baldavia* to the Straits of *Magellan* there are no Towns; the Capes of note are, *Cabo de las Islas*, *Punta del Gada*; and at the very entrance of the Strait, *Cape de la Victoria*, so called from *Magellans* Ship first passing this way.

The West-Entrance of the Straits of *Magellan*, is in **Straits of** 53 degrees of South Latitude; and the East-Entrance **Magellan.** lies in 52 deg. 30 m. the length 110 leagues, and the breadth in some places two leagues over, in others not fully two miles. This place was first discovered and passed through by *Ferdinando Magellan* a *Portugal*; followed by Sir *Francis Drake*; afterwards it grew familiar to many Seamen.

There is another Passage betwixt the South-Sea and **Fretum** Atlantick-Ocean, to the southward hereof, called *Fre-* **le Maire.** *tum le Maire*, found out, Anno 1615, by *Jacob le Maire*, and *William Cornelison Sabouten*, much more convenient than the former; betwixt both which the Land is called *Terra del Fuego*, the South Point whereof is *Cape Horn*: the two Lands betwixt which they sayled when first discovered, they called *States-land* on the East, and on the West *Mauritius-land*.

The next places of note on the main Continent, beginning at *Cape Virgines*, lying at the very easternmost part of the Strait *Magellan*, are *Rio de la Cruz*, where *Magellan* stayed two months.

Rio St. Julian, *Port Desire*, *Rio de las Camarones Cape Rotundo*, *Cape St. Antonio*, at the Mouth of *Rio de la Plata*, a large River, and of so violent a stream, that the Sea, for many leagues together, altereth not its taste.

On the North hereof is *Brasil*, possessed chiefly by **Brasil.** the *Portugals*: a Countrey abounding with exceeding plenty of the best Sugars; that and the great quantity of Red-Wood used for the dying of Cloth, being the chief Commodities hereof.

The places of note, are *Sanctos*, *St. Vincente*, *Saint Sebastian*, at the Mouth of *Rio Janeiro*, *Spirito Sancto*, *Porto Seguro*, *Todos los Sanctos*, *Salvador*, *Olinda* on the River *Maragnan*, *Cape Blanco*, *Pernambuco*, and *Augustine* the easternmost part of *America*.

Paraiba, on the River so called; *Rio de Grand*, *Para*, *Rio de Amazones*, a River full of Islands at the entrance, broad and of a long course, the discoveries whereof are not fully made.

Places to the northward are *Caripa*, memorable for a Colony of the English there planted by Captain *Robert Harcourt*, 1608. on the Bank of *Wiapoco*.

The River *Ormoque*, and *Surenam*, on a River so **Guiana.** called, in the Countrey of *Guiana*, not long since a thriving Plantation of the English, lately delivered into the hands of the Dutch, yeelding Sugars, Cottoni, Tobacco, Wood for Dyers, and some other Commodities.

St. Thoma, the onely Town of *Guiana* inhabited by the Spaniard.

Porto de Guero, *Puerto la Cabela*, *St. Martha*, on the Shores of the Ocean, neighboured by a safe and convenient Haven: spoyled by Sir *Francis Drake*.

Rio de La Hacha, *New Salamanca*, *Sevilla Chez de Mexas*, near the confluence of the Rivers *St. Martha* and *Magdalena*.

Carthagena, situate in a Peninsula; well fortified since the taking thereof by Sir *Francis Drake*, who anno 1585. took it by assault, and carried from thence, be-

B   sides

sides inestimable sums of money, 240 Brass Pieces of Ordinance.

Next *Darien*, near the Strait of Land so called, on the Bank of the River *Noelte*.

Nombre de Dios. *Nombre de Dios*, conveniently seated on the upper Sea for a Town of Trade, whither the Spaniards brought their Goods from *Spain* for *Panamo*, and from *Panama* for *Spain*; taken also by Sir *Francis Drake*.

St. *Philips*, situate on a safe and strong Haven called *Porto Bell*, built in this place by the command of King *Philip* the Second, to be the Staple of Trade betwixt *Spain* and *Panama*, instead of *Nombre de Dios*, where it was before; removed partly because of the unwholsome Air of *Nombre de Dios*, but chiefly because that Town lay too open to the English Invasions; fortified with two strong Castles on each side of the Haven, yet for all that, Surprized and Pillaged by Captain *Parker*, Anno 1601.

Gulf of Mexico. *Baya de Cartago*, *Cape de Honduras*, *Porto de Sol*, *Porto de Cavallos* the most noted Haven in the Gulf of *Honduras*; whence compassing the Peninsula of *Yucatan* by *Cape de Cotoche*, the great Bay or Gulf of Mexico openeth at self; the Ports and Places of the Shores whereof were heretofore little frequented, unless by the Spaniard; in these later times, and since the Plantation of *Jamaica* by the English, something better, though yet not much known.

The chief, and almost the only place, is *Lavera Cruz*, the next Port Town to the great City of *Mexico*, from which it is distant about sixty leagues.

The Traffick and Commodities of these Eastern parts of *America* being the same with those on the Shores and Coasts of *Mar del Zur*, spoken of before.

North-eastward hereof lyeth the Coast of *Florida*; betwixt which, and the Coasts of *Guiana*, before treated of, lie scattered up and down the Sea a great number of Islands, some greater, some lesser, *viz.*

Cuba. *Cuba*, *Hispaniola*, *Jamaica*, *Porto Rico*, the *Caribee*, and *Lucayes*.

Cuba a large Island, in length from *Cape Maysi* near *Hispaniola*, to *Cape St. Antonio*, 230 leagues; a fertile Soil, liberally stored with Ginger, *Cassia*, Mastick, Aloes, Sugar, &c.

Ports of most note, St. *Jago*, *Salvador*, *Santa Cruz*, *Santa Spirito*, *Trinidad*, *Port del Principe*, *Baroca*, *Matanca*, and the *Havana*, a noted and well-traded Port, so strongly situate and fortified, both by Nature and Art, that it seems impregnable.

Hispaniola. *Hispaniola*, a large Island also, but not so big as *Cuba*, a plentiful and pleasant Country, once abounding in Gold, but long since exhausted; it affords Ginger and Sugar in abundance.

The Ports worthy observation, St. *Domingo*, the Residence of the Governor, not yet recovered of the Damage done by Sir *Francis Drake*, St. *Salvador*, *Ingoana*, or *Santa Maria del Porto*, *Porto de la Plata*, *Agua* or *Compostela*, &c.

Jamaica. *Jamaica*, on the South of *Cuba*, from whence distant twenty leagues or thereabouts; and not much more from *Hispaniola*; formerly possessed by the Spaniard, not many years ago taken by the English, who therein have began a gallant Plantation; the wholsomness of the Air, and fertility of the Soyl, giving great hopes (if not assurance) of a continued encrease and improvement thereof, to the encouragement of such as are already there, or others that shall hereafter transport themselves thither. Merchandize of their own growth, are Tobacco, Sugar, Cotton, Ginger, Indigo, and several sorts of Woods serviceable for Dyers and others, Places of note are *Sevilla*, *Melilla*, *Gresian*, *Punta Nigrilla*, *Port Royal*, *Port Morante*, *Aguia*, &c.

Porto Rico. *Porto Rico*, something Mountainous, but indifferent fruitful, exposed sometimes to those sudden and troublesome Tempests, called Hurricanes, as are the rest of these places hereabout. The Commodities, Ginger, Sugar, Cassia, and Hydes; the European Cartel so encreasing in most of these Islands, that they have grown wild by reason of their multitude; the inhabitants of many of the places killing thousands for their Skins only, leaving their flesh as a prey to ravenous Creatures. Places of note are, *Porto Rico*, *Arecibo*, *Loyza*, &c.

Caribbe Islands. The *Caribes*, or *Cannibal* Islands, so called in general, because at first discovery inhabited by *Cannibals*, or Man-eating people, as the word imports; extended in the Sea like a Bow, of different temper and quality; the principal are these, *Margerita*, *Trinidada*, *Grananilla*, St. *Lucies*, St. *Vincent*, *Barbados*, a flourishing Colonie and Plantation of the English, well peopled, the Soil in shew like *England* but more fruitful; furnished on the South side with a large and commodious Haven, driving a great Trade in Tobacco, Sugar, Cotton, Ginger, Indigo, and Logwood, &c.

Next *Martinico*, *Dominico*, *Mary-galant*, *Desseda*, *Guardalupe*, *Antego*, *Barbada*, *Mount Serat*, St. *Christophers*, *Nevis*, St. *Martins*, St. *Bartholomew*, *Anguilla*, *Santa Cruz*, and many others of less note.

Lucaioes. The *Lucaies* are *Majaguana*, *Samana*, *Yuneto*, *Yama*, *Guanahani*, *Cygnateo*, *Lucayaque* and *Bahama*, memorable for giving name to the violent Current interposing betwixt it and the deny Island of *Florida*, of so forcible a course, that no strength of Wind or Oars can prevail against it, (as is commonly reported).

Florida. *Florida* was first discovered by the English, under the command of *Sebastian Cabot*, Anno 1497. so called by *John de Ponte*, afterwards from the fresh verdure and flourishing estate in which he found it: The Ports are *Sancta Lucia*, St. *Augustine*, St. *Matthews*, *Port Royal*, *Cape Feare*, *Port Charles*, and St. *Helens*, which three lie near the borders of *Virginia*.

Virginia. *Virginia*, a gallant Plantation of the English, having many excellent properties above other Nations, as the temperature of the Air, fruitfulness of the Soyl, commodiousness of situation; many great and navigable Rivers, and safe and spacious Harbours. The first discovery hereof by the two *Cabots*, Father and Son, Anno 1497. did first entitle the Crown of *England* to this Country, who still possess it, having there a large and flourishing Plantation. The chief Trade, besides other Commodities, is Tobacco, where there is such abundance, that no place affordeth more, or of better quality. The Rivers are, *James River*, *York River*, *Potomac*, *Rapahanock*, *Elizabeth River*, *Warwoneco*, and many others, all falling into the great Bay called *Chesapeach*.

The two Capes, at the entrance whereof are Cape *Henry*, and Cape *Charles*. Towns of most note are, *James*-Town, the Seat of the Governour, and many others.

Eastward of *Virginia* lyeth the Isles of *Bermudes*, so called from *John Bermudas* a Spaniard, by whom it was first discovered: Also called the *Summer* Islands, from Sir *George Summer*, who there suffered shipwrack; there are several of them, altogether making a body in form of a Cressent, and inclose very good Ports, as those of *Southampton*, *Harrington*, and *Pagets*. The Air is almost always serene, very healthful, agreeing well with English bodies, who have here at divers times settled and established a fair and powerful Colonie. The healthfulness of the place inviting that famous Mathematitian Mr. *Richard Norwood*, once Reader of *Gresham* Colledge to *London*, to make his abode here. Cochineel and Tobacco, with some Pearls, Amber, and fair Oranges being their principal Riches, for which they have a good Trade.

Mary-land. To the North-east of *Virginia* lyeth *Mary-land*, and *New-England*, a Country bravely situate, and very agreeable

greeable to English bodies. The Soil exceeding fruitful of Natures necessities, even to excess; supplied also with many large and capacious Bayes and Rivers. The Commodities, besides store of Flesh and Corn sent abroad, are Furs, Amber, Flax, Hemp, Cedar, Pitch, Tar, Masts, Cables, and Timber for Shipping and other uses; in a word, whatsoever comes to England from the Sound, might be as well supplied from hence.

New England The chiefest Places are New-York, seated on the great River Manhattans, or Hudsons River, near its fall into the Ocean, and not far from the Isle Mattenacks, or Long-Island, over against the East end whereof the River Connecticut falleth into the Sea.

The next Boston, Barnstaple, New-Plimouth, near unto which is that observable hooked Point of Land named Cape Cod, with several others: And St. Georges Fort, built by the English, at the Mouth of the River Sagahadar.

Nova Scotia. Adjoyning hereunto lyeth Nova Scotia, Nova Francia, or Canada, and the small Peninsula, called Accadia, between the Bay of St. Lawrence, where the great River Canada falleth into the Gulf and the main Sea. The noted places are Port Royal, St. Lukes, Port au Mouton, Gaspe, Gachepe, St. Croix, Francis Roy, and St. Laurence. The Commodities are chiefly Furs, &c.

Not far to the eastward hereof, lyeth the Isle of Sables.

New-found-Land. At the most Eastern part of America Septentrionalis, lyeth New-found-land, an Island separated from the Main, or Terra Corterialis, by a Frith or Strait, called Gulf de Cosseaux, furnished on the Sea-Coast with abundance of Cod-fish, and other Fish; for the catching of which, Ships of many Countreys frequent that place, having also many large and convenient Havens, as Renvesse, Fass-Haven, Thorn-Bay, Trinity-Bay, Bonavist, White-Bay, Port Tressass, St. Georges Bay, St. Jones, &c. The Air of this Countrey never very extream, more temperate in the depth of Winter than with us in England, the Brooks being never so frozen over that the Ice is able to bear a Dog; and those little Frosts but seldom holding three nights together.

Before the Island lyeth that long Bank, extending in length some hundred of leagues; near to which are many little Islands, called by John Cabot, Baccalor, peculiar now to one onely, from the numerous multitude of Cod-fish which swarmed hereabout. Betwixt Cape de Gamay in Terra Corterialis, and the Cape Farewel and Desolation, near Greenland, lyeth the two Straits, named Fretum Davis, and Fretum Hudson, so called from the first Discoverers; a Sea diluting it self much both towards the North, South, and West, giving great Hopes thereby of a Passage to China, and the East-Indies: And therefore notwithstanding the Ice, Fogs, and other Incumbrances searched into by many English Worthies, as Frobisher, Davis, Weymouth, Hudson, Button, Baffin, Smith, James, Gillam, and others, who have sailed therein, some one way, and some another, and given names to many places, viz. King James his Cape, Queen-Ann's Cape, Prince Henry's Foreland, Saddel Island, Barren Island, Redgoose Island, Digs his Island, Hackluyts Headland, Smiths Bay, Prince Ruperts River, Mauditn Savad, Fair-haven, and many others, even from James his Bay on the South, at the bottom whereof Hudson wintered in the latitude of 51 degrees, to Baffins Bay on the North, lying in the latitude of 79 degrees; and to the westward, as far as Port Nelson, where Sir Thomas Button wintered, being more westerly than Mr. Hudsons Bay by 190 leagues; and near as far towards the West as Cape California in Mar del Zur, where finding the Tyde continually to rise every twelve hours fifteen foot or more, and that a West Wind did make

Freti un Davis. Fretum Hudson.

the Neap Tydes equal to the Spring Tydes, finding also the Tydes to set sometimes eastwards, sometimes westwards, gave good hope to Mr. Hubard (who made the Plat thereof) of a through Passage, called therefore Hubards Hope.

And in the year 1667, a very honourable and worthy Design was renewed, and undertaken for the discovery of this North-West Passage, and setling a Trade with the Indians in those Parts, by several of the Nobility of England, and divers Merchants of note belonging to the City of London, who fitted out two small Vessels for that purpose, the one called the Nonsuch Ketch, Captain Zachariah Gillam Commander, the other the Eaglet Ketch, Captain Stannard Commander; the latter whereof being by Stormy Weather beaten back, returned home without success; but the other proceeding on her Voyage, in her way made the Land of Bass, lying betwixt Iseland and Greenland; passed through Hudsons Straits, then into Baffins Bay; from thence southerly into the Great Bay, where in the latitude of fifty one degrees, or thereabouts, in a River now called Ruperts River, he wintered, found a friendly correspondence and civil entertainment with the Natives, traded with them in exchange of Bever-Skins, &c. for Knives, Beads, Looking-Glasses, Hatchets, and other trivial Commodities, and the next year returned with good success, and future hopes of an excellent Trade in those parts, giving invitation thereby to the aforesaid Noblemen and Merchants to adventure again, Anno 1669. Which Voyage being not yet performed, leaves us ignorant both of the Places and Trade thereof (save what is already known) undiscover'd, till the conclusion of the Voyage.

Greenland. Greenland, called by the Natives Secannunga, is that last part of America, which remains onely to be spoken of; a Countrey, as is supposed, but chiefly inhabited, and by reason of the abundance of Ice, and inhospitableness of the People, little frequented, and consequently not so well known, as to give a perfect description thereof: for notwithstanding several Voyages have been made thither on purpose, many Ships have accidentally touched upon the Coasts thereof in pursuance of the N. W. Discoveries; yet for the coasts aforesaid, the Countrey lies still obscured in a Northern Mist, being to us almost altogether unknown, unless the names of certain Bayes, Capes, and such like, as Whale-sound, Horn-sound, Rensbels-sord, Cunningham, Gilberts-sound, Cocking-sord, Cape Comfort, Cape Disford, Leister-point, Warwick-Foreland, Harroldts-Ness, Bereford, not far from the West part of Iseland, and several others.

Groenland. South-westward from Iseland, about 140 leagues, lyeth an Island called Gust, in the latitude of 57 degrees 35 minutes, not yet fully discovered; but only as it hath been accidentally seen by some, who upon other Discoveries have occasionally passed those Seas, as Captain Gillam in his first Voyage to the North-West Passage had Soundings near unto it.

Trinity Island. From Iseland, about 135 leagues North-eastwards, in the latitude of seventy one degrees, lyeth an Island called Trinity Island, the North-East Point whereof is named Young-Foreland, a place formerly much frequented by the Dutch for their Whale-fishing; the Land is very high, full of Rocks and Mountains, one especially much higher than the rest, called Beara Mountain.

Thus briefly have I touched at, and as it were, onely named the Sea-Coasts of most of the known Parts of the World, which may serve as an entrance to the succeeding Sea-Atlas, commending the Discovery of such parts as are yet unknown to the search of Posterity.

The Discoveries that have been made within this two hundred years, by the Worthies of our own Nation, as well as Strangers.

Christopher Columbus the Genuer, first determinately attempted to seek after, and in the year 1492 prosperously discovered the great Continent of America.

John Cabot a Venetian, and his Son Sir Sebastian, born in England, succeeded Columbus in that famous Attempt, and discovered all the North Coasts thereof, from Cape Florida to New-found-Land.

But Americus Vespucius, in discovering some of the South Parts thereof, obtained the honour of having the whole Continent called by his Name America.

Ferdinando Magellan, Anno 1519, was the first that found out that Strait towards the Antartick Pole, which gives a Passage between Mar del Zur and the Atlantick Ocean, called by his Name, Fretum Magellanicum.

Mr. Richard Chancelor first found out the Passage by Sea to Russia, Anno 1550.

Sir Hugh Willoughby first discovered Greenland, or King James his New-Land; attempted to find the North-East Passage to Cathay and China, Anno 1553, but in his return was frozen to death.

Mr. Stephen Burroughs attempted the like Passage, anno 1556. discovered several places in Russia, Nova Zembla, and thereabout, to his lasting memory.

Sir Francis Drake, that adventurous and valiant English Worthy, after a Voyage first made to Nombre de Dios, and other parts of the West Indies, in the years 1572, and 1573, having then only a sight of the South Sea, renewed in himself a noble desire of sailing therein; and after some hindrances at home, in Service of his Prince and Countrey, anno 1577, by gracious Commission from his Sovereign, and the help of divers Friends, Adventurers, fitted himself with five Ships for his intended design, and passing through the Straits of Magellan, made many rare Discoveries there, and on the West of America; sailed thence through the South Sea to the East Indies, and steering homeward by the Cape of Good Hope, after two years and ten months spent in that Circum-navigation of the World, and many excellents Atchievements and Discoveries there performed, that worthy Knight, and most noble Neptune, happily returned to Plymouth whence he first set forth. Other Voyages he made sometime afterwards to the main Continent of America, and the Islands thereof, wherein were taken by that English Hero, the City of St. Jago, Sancto Domingo, Cartagena, with the Fort and Town of St. Augustine in Florida.

Mr. Tho. Cavendish not long after followed the trace of Sir Francis through the Straits of Magellan, encompassed the whole circumference of the Terrestrial Globe, in the year 1587, and prosperously returned (laden with honour and applause) into his Native Countrey.

Several famous Men among the Netherlanders have also, to their lasting credit, encompassed this Globe of Earth and Sea, as Oliver van de Nort, Isaac le Maire, and William Cornelison Sebouten, who according to their several Courses and Voyages, made discoveries not to be forgotten by Posterity.

Sir Martin Frobisher, Anno 1576, attempted the North-West Passage, sailed to the latitude of 62 degrees, found that great Inlet, since known by his Name, Frobishers Strait.

Mr. Arthur Pett, and Mr. Charles Jackman, Anno 1580, went out in two Ships for the discovery of the River Ob, and a Passage to China, arrived at Vaigats, passed the Straits, took particular observation along the East Part of Nova Zembla, the North of Russia and Samoeds Countrey, so far as the Ice would give him leave.

Sir Humphrey Gilbert, Anno 1583, going for Discovery of the North of America, came into the great River St. Laurence in Canada, settled the Government of the Fishing there.

Master John Davis attempted the discovery of the North-West Passage, anno 1585 came into the latitude of 66 degrees, plyed alongst the Coast, observed the probability of a Passage there, and so the end of the year returned.

The next year went again for a further Discovery, found a great Inlet betwixt the latitude of 55 and 56 degrees; Traded with the People of the Place, and so returned.

In the year 1587, he took a third Voyage for discovery of those Parts, followed his course to the North and North-West, as far as the latitude of 76 degrees, having the Continent on the West, and Greenland (which he named Desolation) on the East; and passing on to the latitude of 86 degrees, the Passage enlarged it self so, that he could not see the Western Shoar; then he altered his course southerly to the latitude of 73 degrees, in a great Sea free from Ice, of an unreasonable depth; and by reason of the departure of two Ships which he left a Fishing, he returned home. This Passage (as he was the first Discoverer) he called by his own name, Fretum Davis.

The Discovery of these Lands, Coasts, Islands, Straits, Havens, Bayes, Rivers, &c. with the Commodities and Advantages arising from the same, in a Treatise of his own, called the Worlds Hydrographical Description, with his yearly Reporteries and Journals, may more largely appear.

Mr. Stephen Bennet first discovered Cherry Island, Anno 1603. at the Charge of Sir Francis Cherry, and therefore beareth his name.

Mr. Henry Hudson, Anno 1608, was sent to discover towards the North Pole, came to the latitude of 81 degrees, attempted the North-East Passage in two Voyages; performed one worthy discovery to the North-West into a great Bay called by his own name Hudsons Bay.

James Pool and Thomas Edge, made a Voyage Northerly toward the Pole, first began the Whale killing in Greenland; to the eastward whereof Mr. Edge found another Island, called by his own name, Edges Island.

F I N I S.

Totius Terrarum Orbis Tabula

NOVISSIMA TOTIUS TERRARUM ORBIS TABULA

A Chart of the
NORTH SEA
By John Seller
Hydrographer to the King

ENGLAND

NORWAY

JUTLAND

SWEDEN

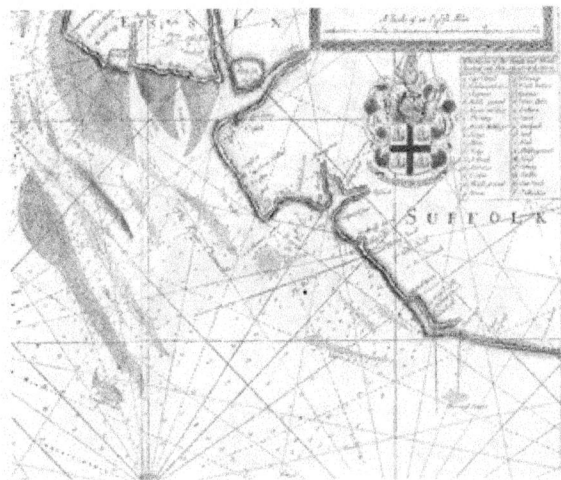

From Yarmouth to Orwell June North Coast of
ENGLAND
East Coast of Scotland

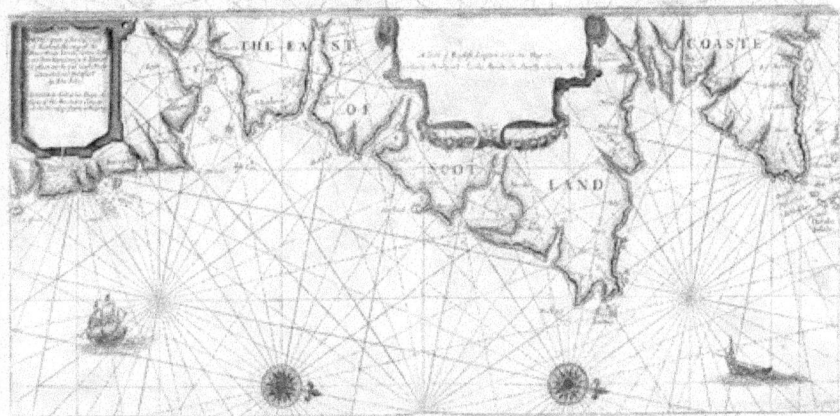

Sea coast of England Flanders & Holland

Flanders

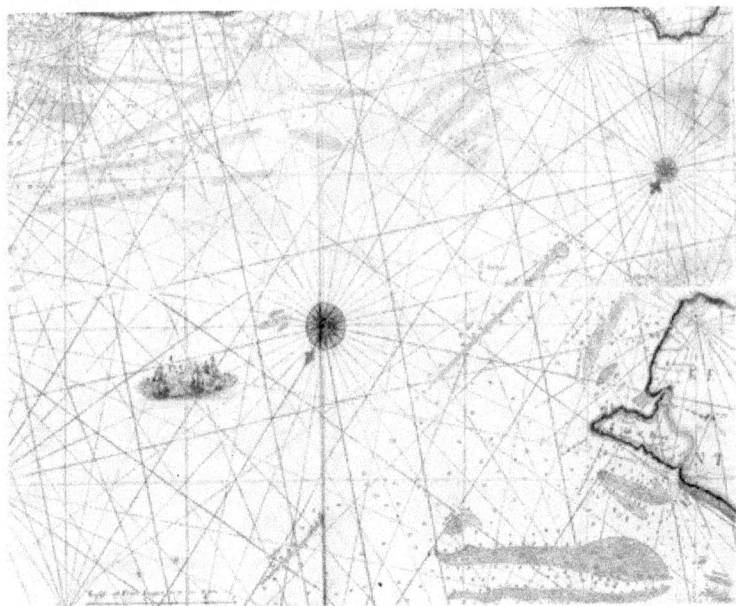

South Sea 2. d. 4. m —— Doues —— Fly. Amsterdar

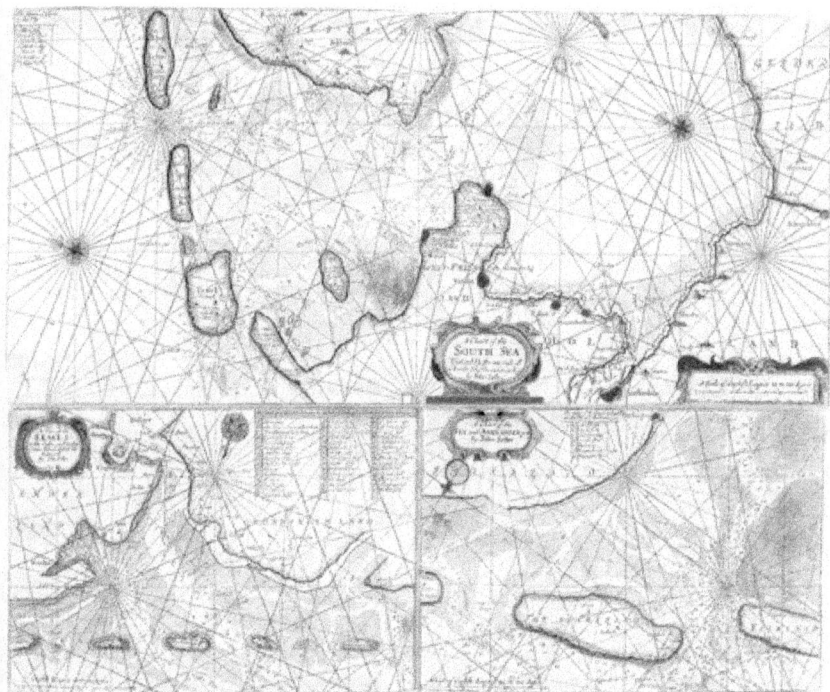

A Chart of the SOUTH SEA

Lewis the — Coast of Scotland &c

Chart of the Maes and Goeree, Coast of Holland.

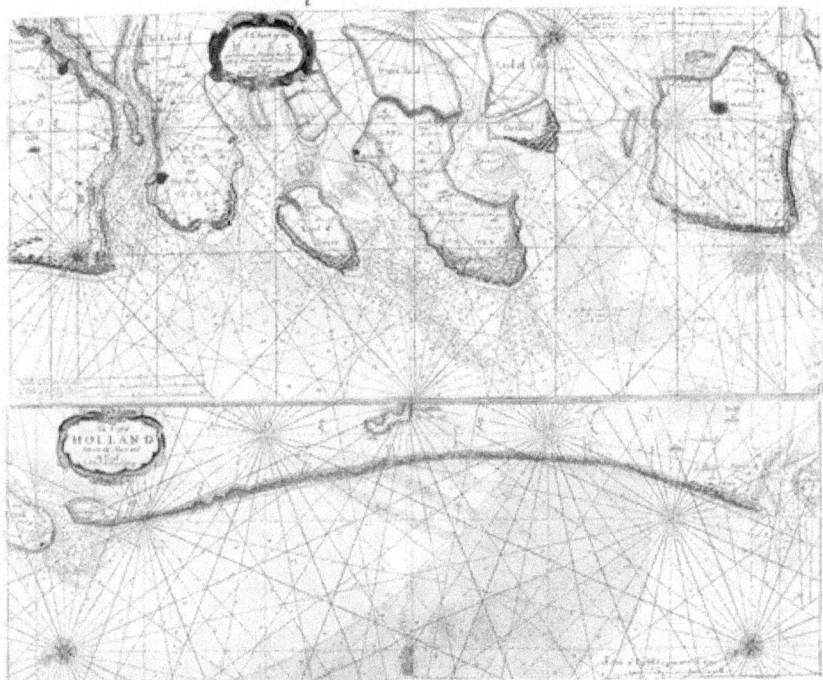

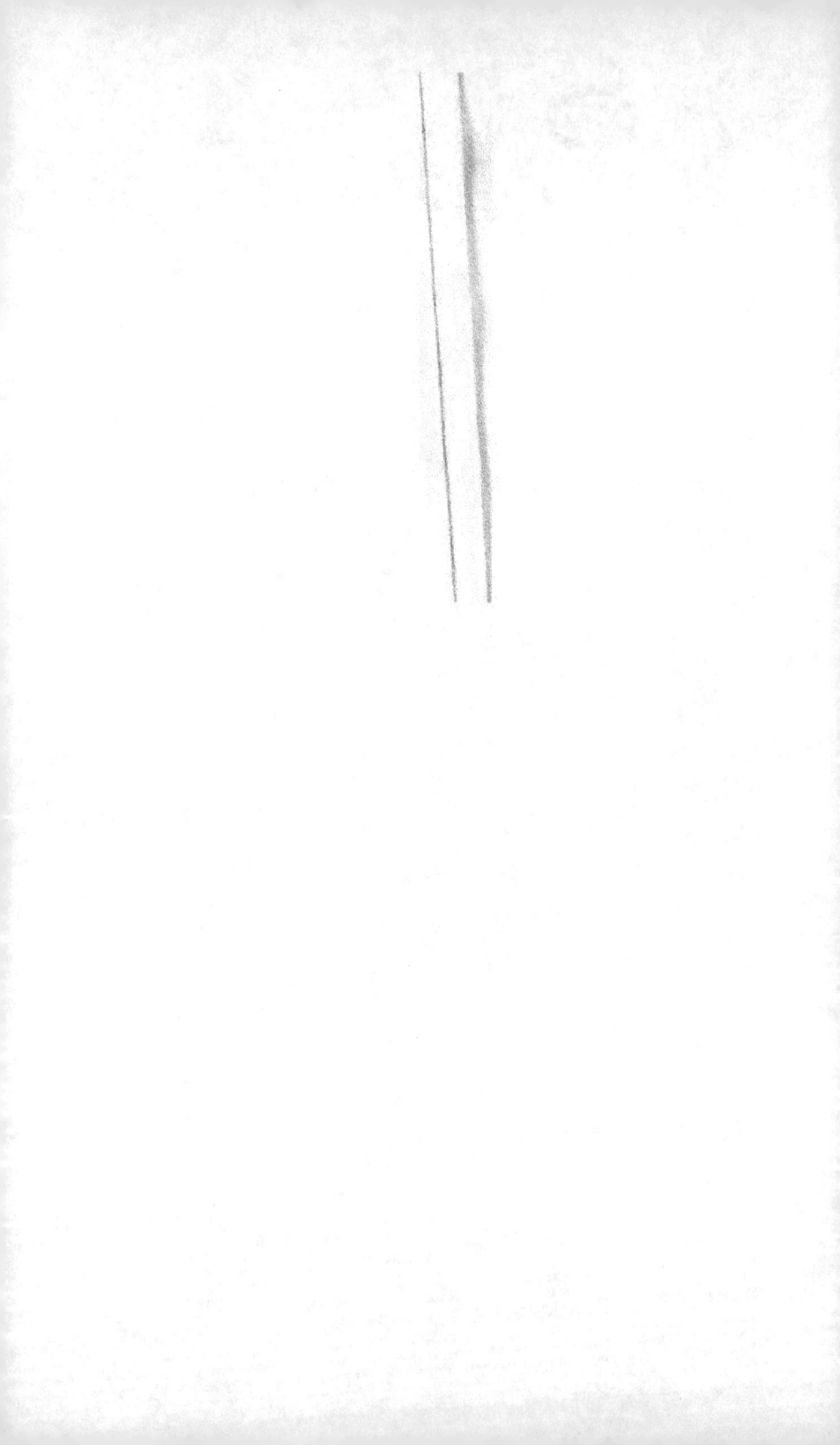

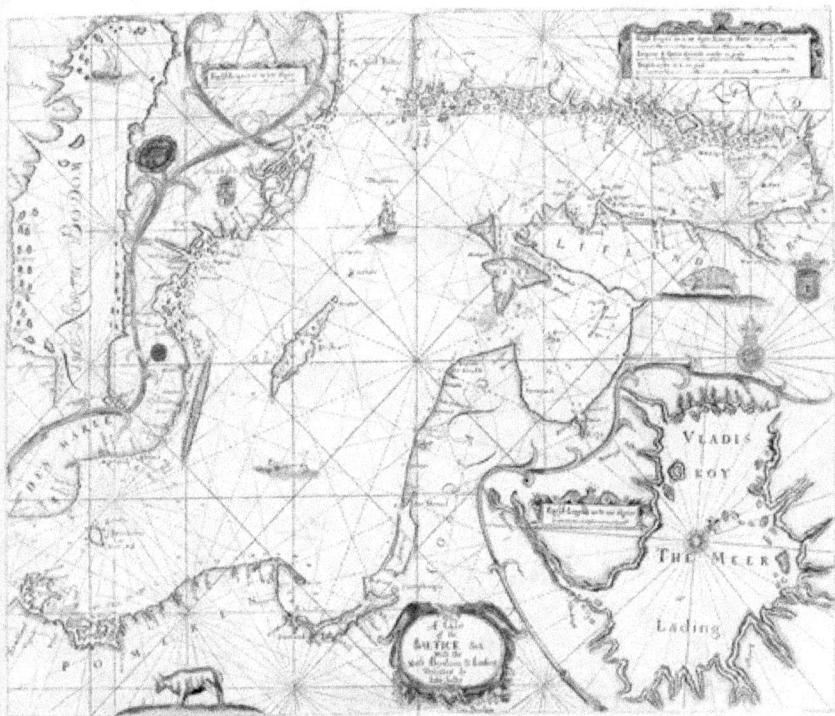

Russia, Lapland, Finmark, Nova Zemla & Greenland

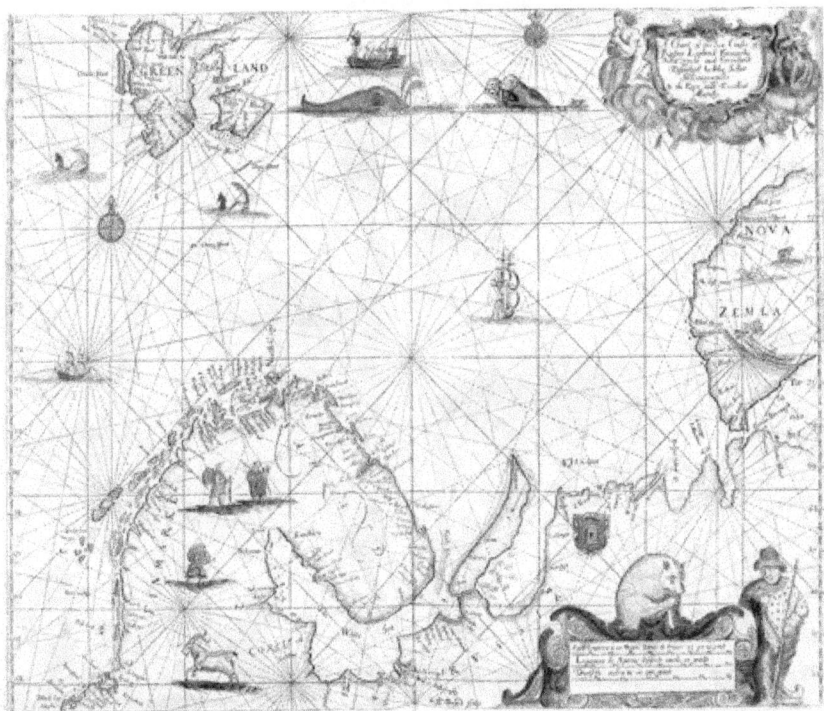

A View ... on the River Dwina going up to Archangel

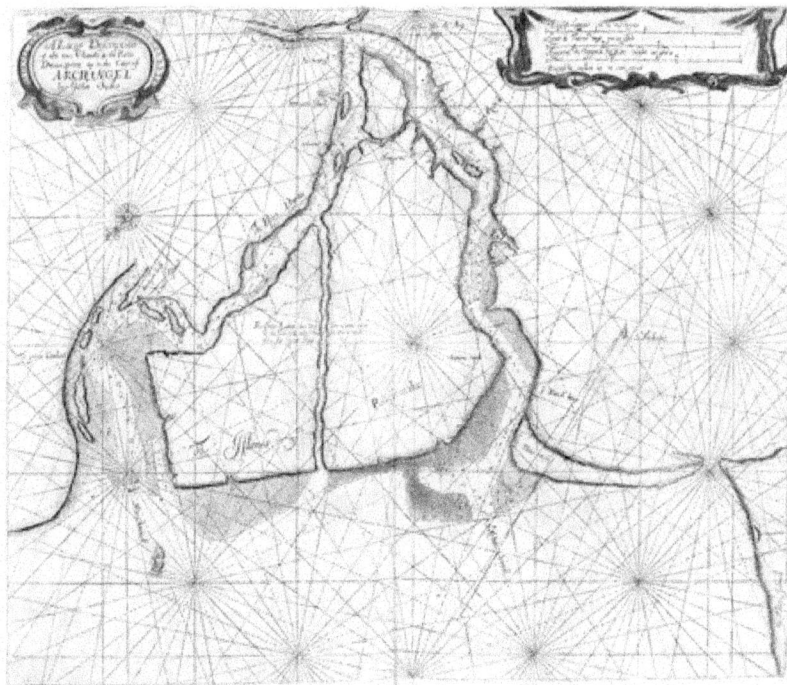

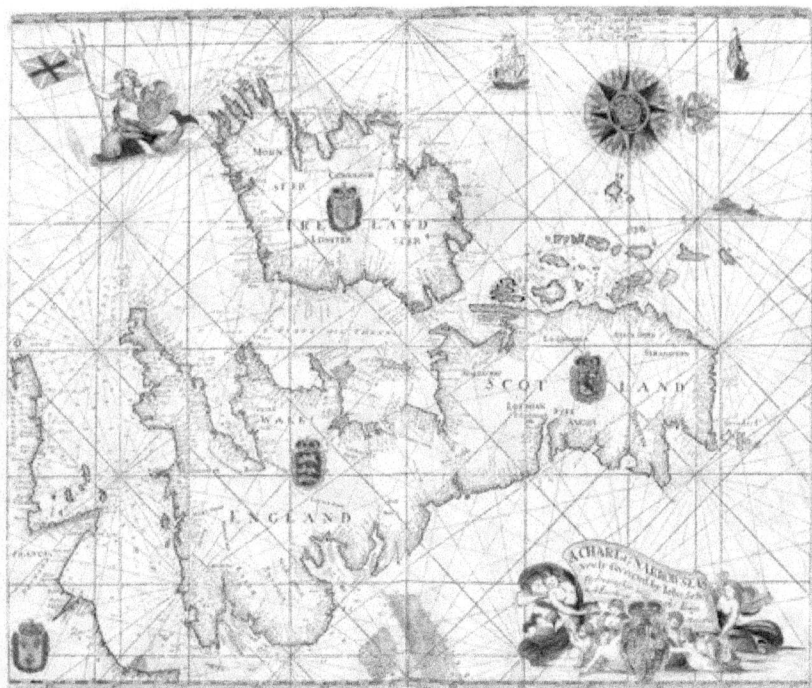

British Ch

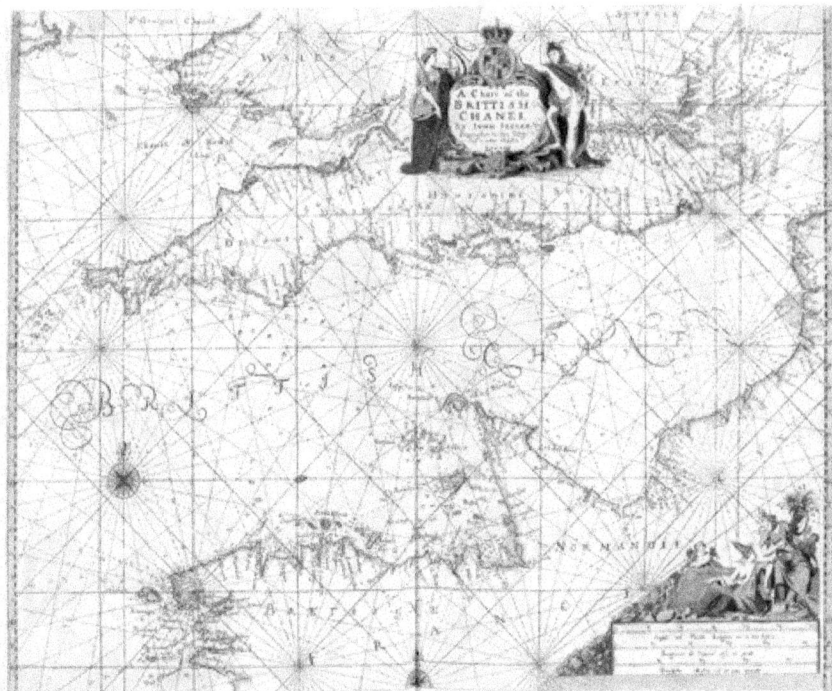

Bay of Panama

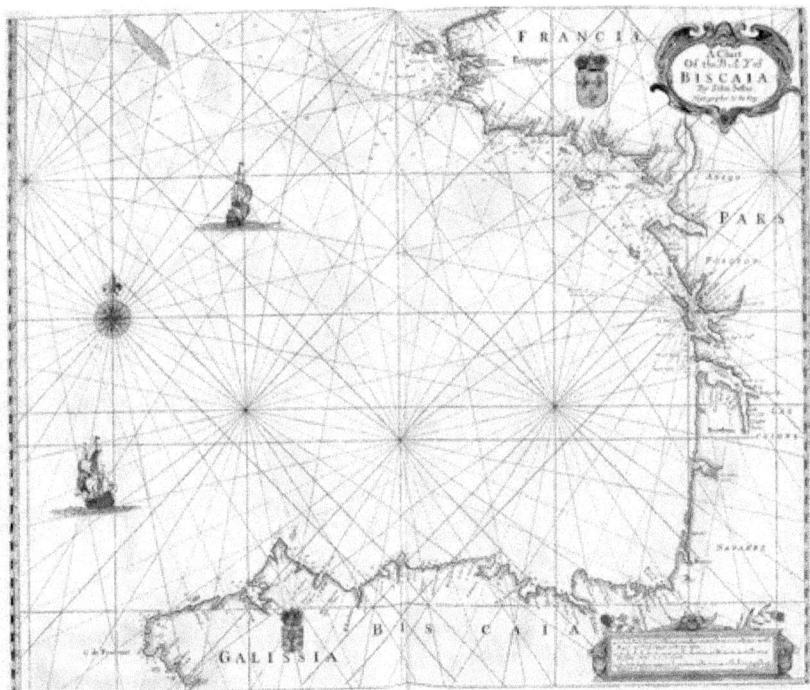

FRANCIE

A Chart
Of the B: L: Y of
BISCAIA
By: John Seller,
Hydrographer to the King

PARS

GALISSIA   BIS CAIA

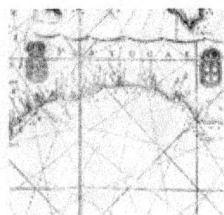

A Chart of the
Westernmost Part of the
MEDITERRANEAN
SEA
By JOHN SELLER
Hydrographer to the King

Levant or Eastern part of the Mediterranean Sea

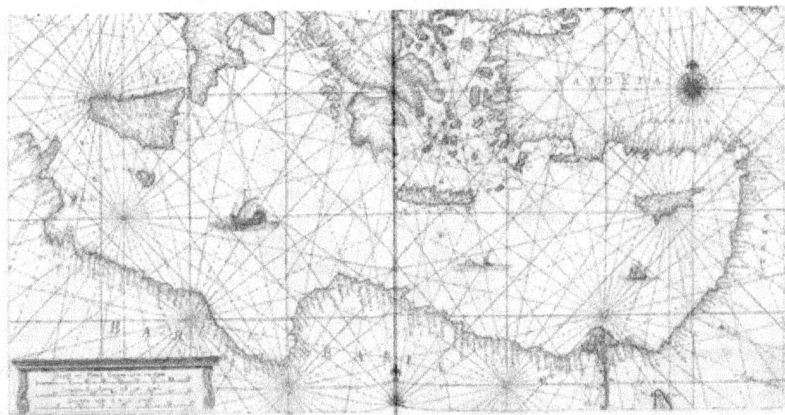

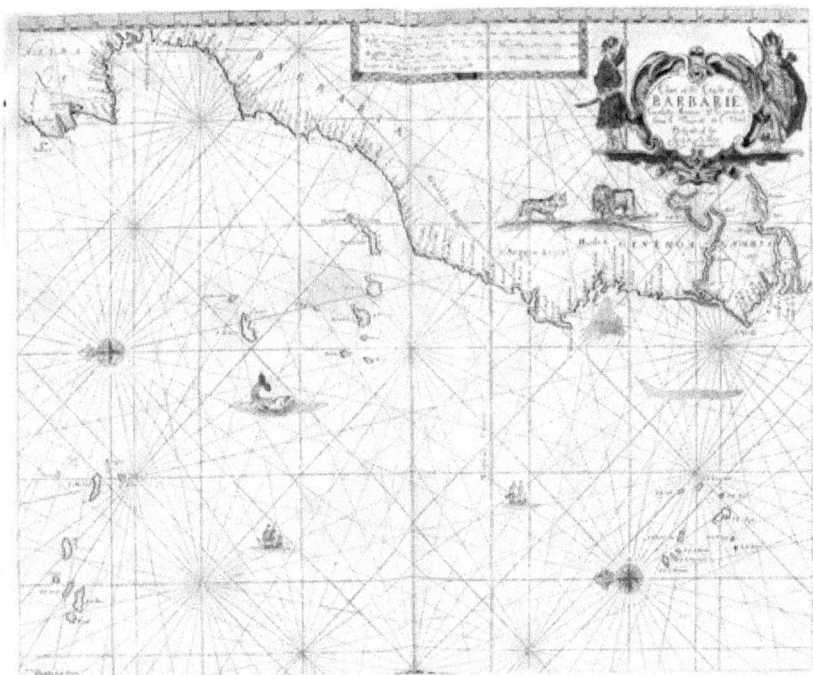

Guinea

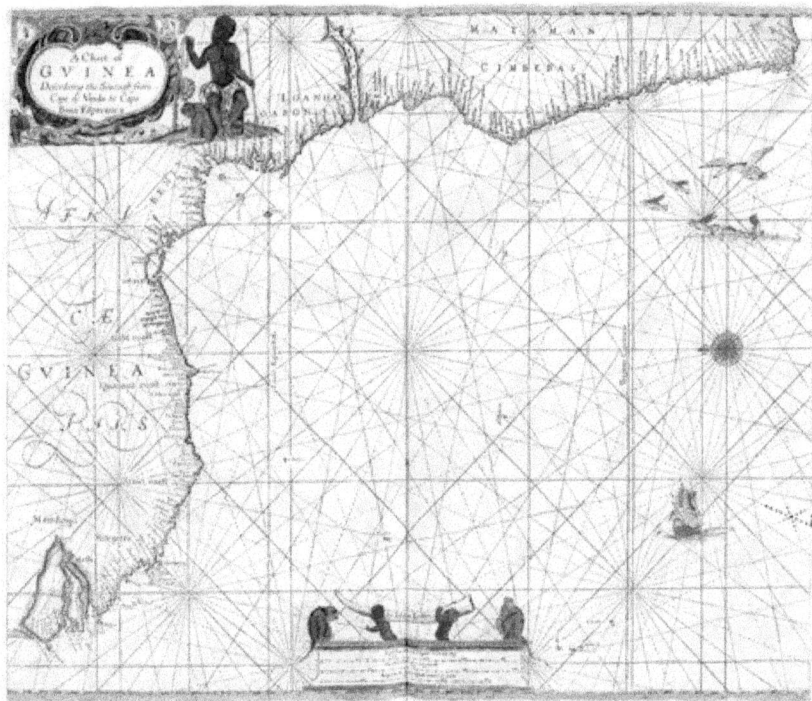

Western part of the East In

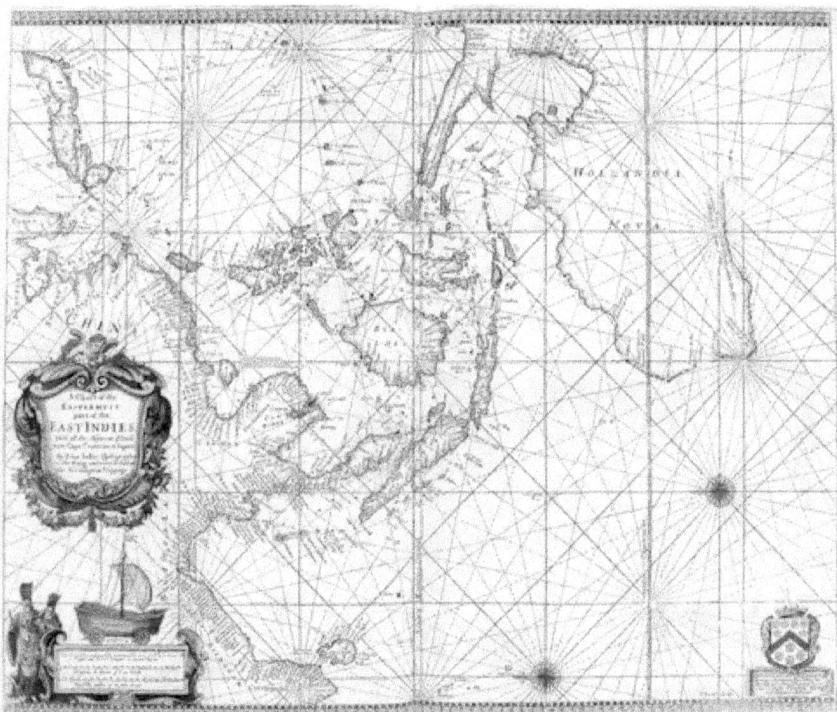

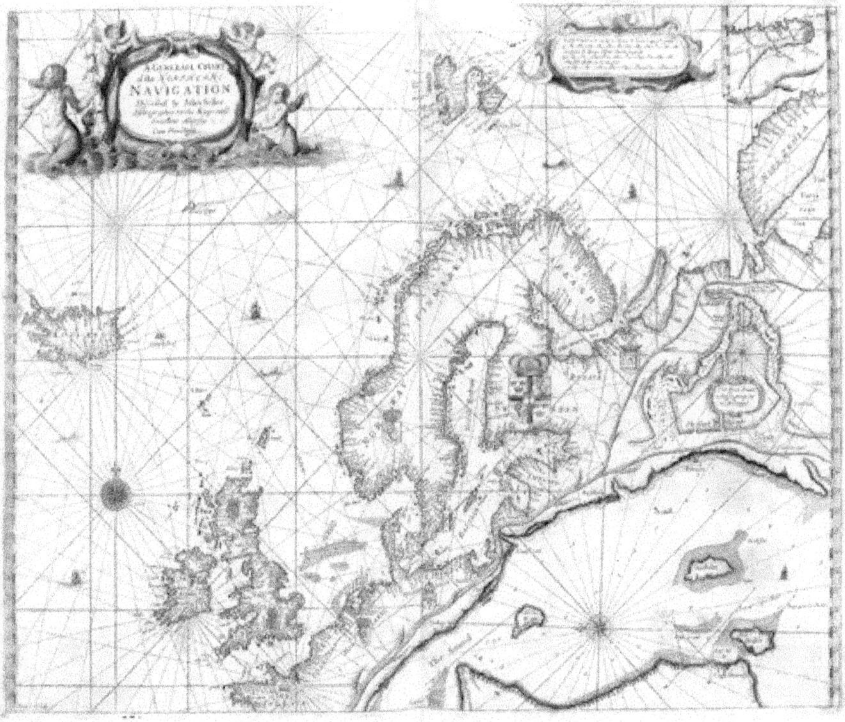
NAVIGATION

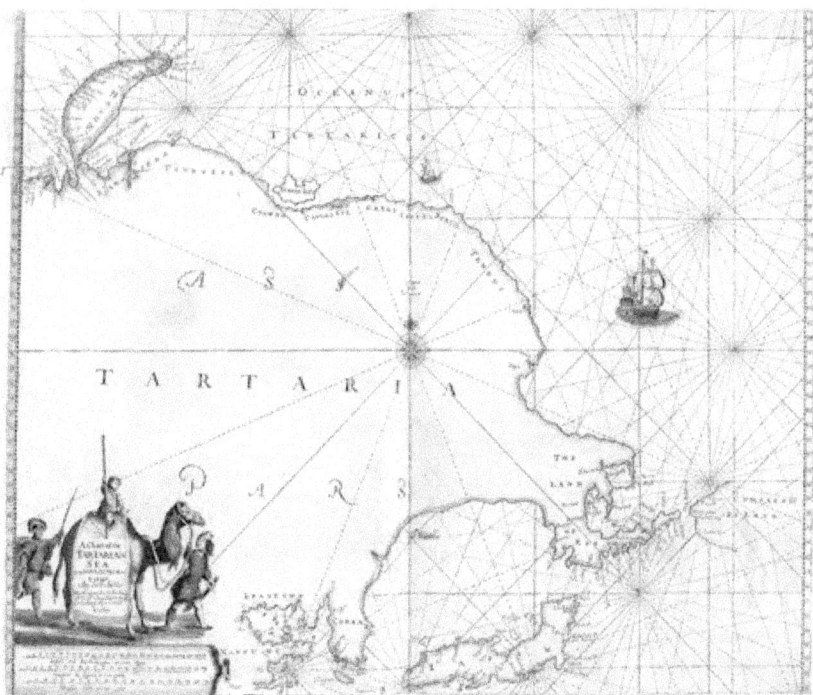

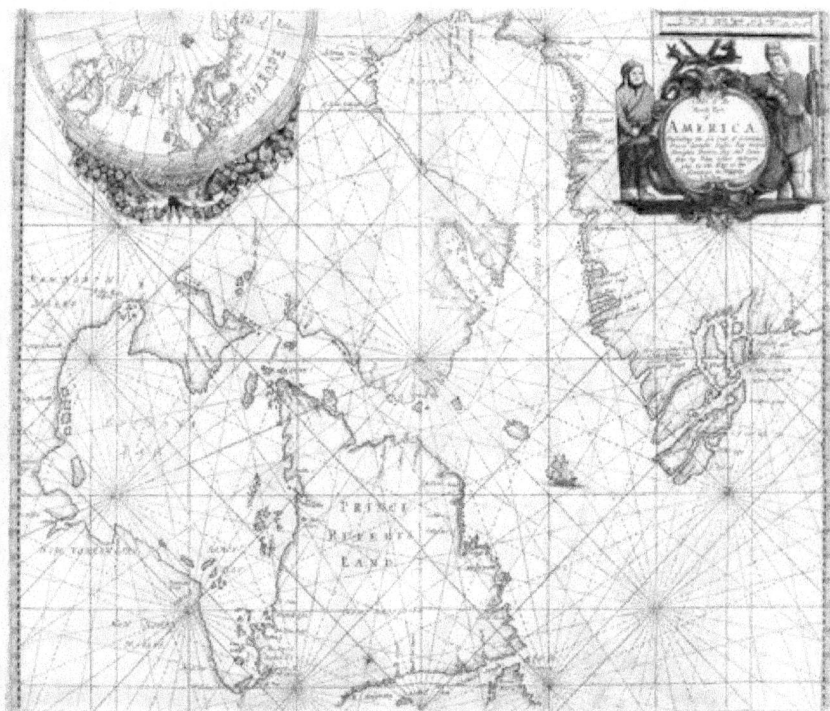

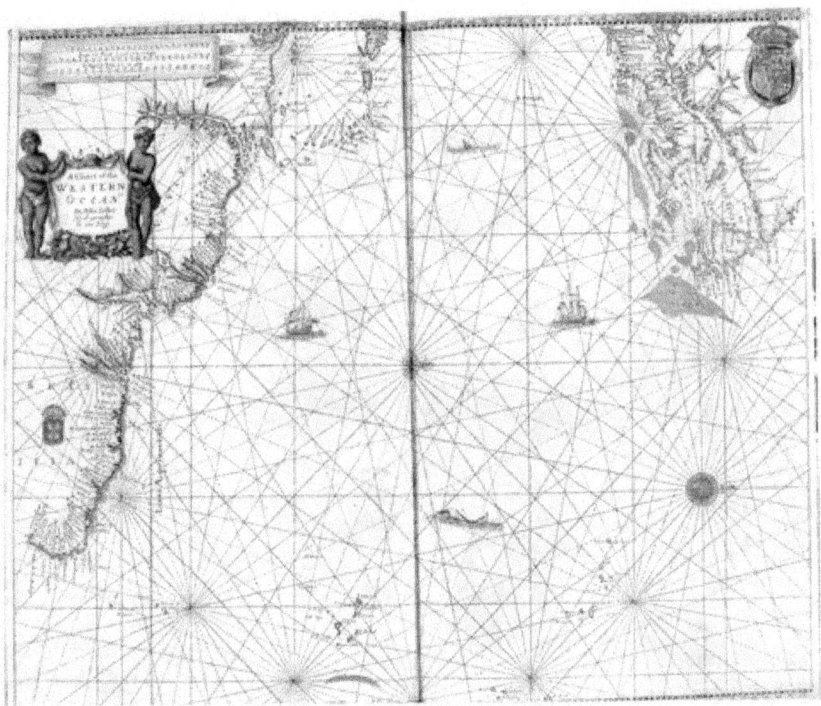

A Chart of the WESTERN OCEAN

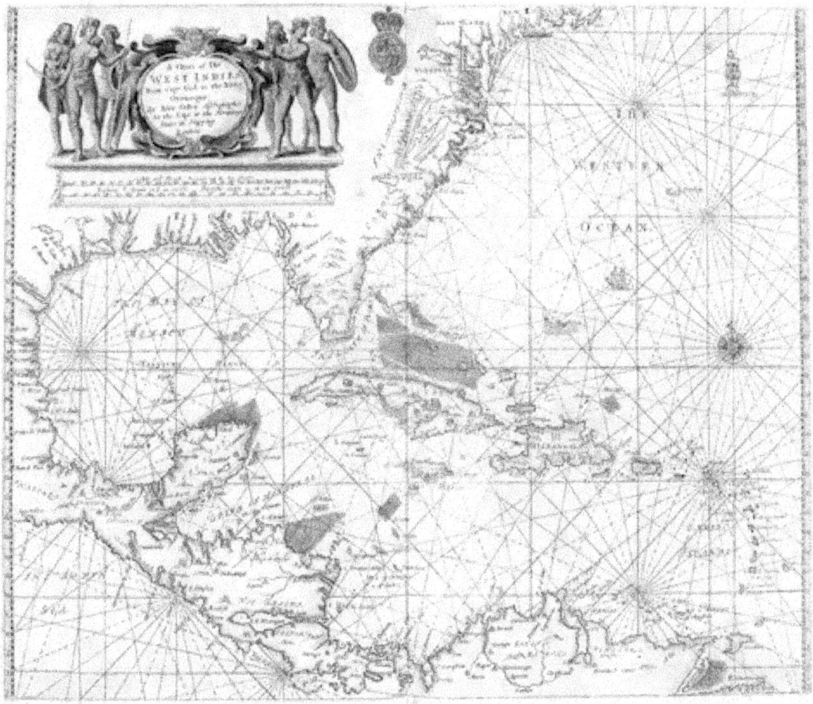
WEST INDIES
OCEAN

Caribe Islands

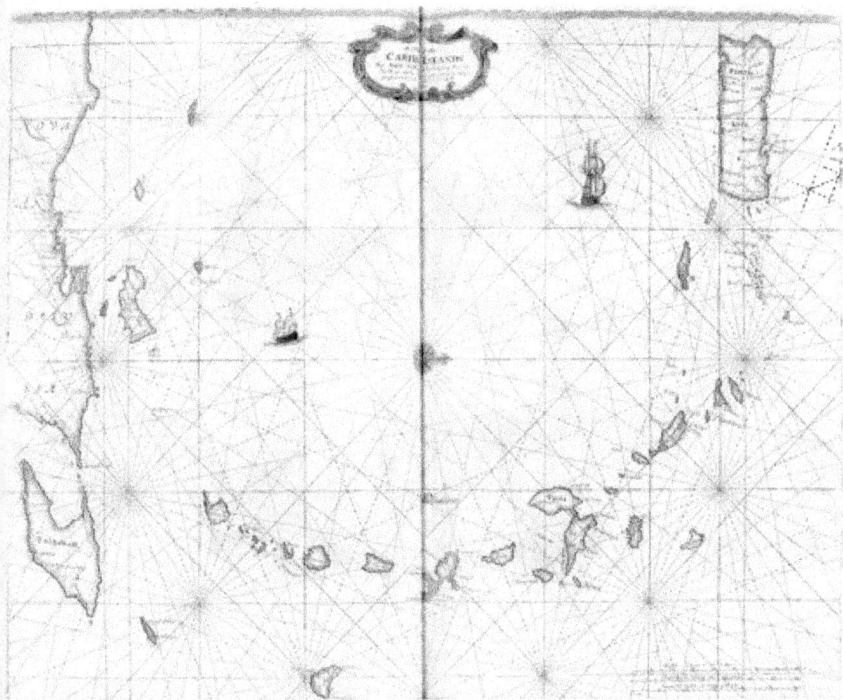

CARIB ISLANDS

Windward Passage, from Jamaica

Based on the handwritten Table of Contents, an inventory in September 2005 determined
that the following maps were missing from this volume:

    Map 28 (New Jersey)
    Map 30 (West Indies from Cape Cod to River Oronoque)

With the maps that were recovered from the Forbes Smiley theft, two similar maps were
returned to the Boston Public Library. Since internal evidence indicates that John
Seller's "A Chart of the West Indies from Cape Cod to the river Oronoque" was
originally bound in this atlas, it has been re-inserted as Map 30 as part of this
conservation treatment. Although Smiley admitted removing "A Mapp of New Jersey in
America" from this atlas, there is insufficient internal evidence to suggest that it was the
one that was originally bound in this atlas. Consequently, we have not re-inserted it as
Map 28 during this conservation treatment. It will be cataloged and filed separately as a
single sheet map.

Ronald E. Grim
1/5/2011

Windward Passage, from Jamaica

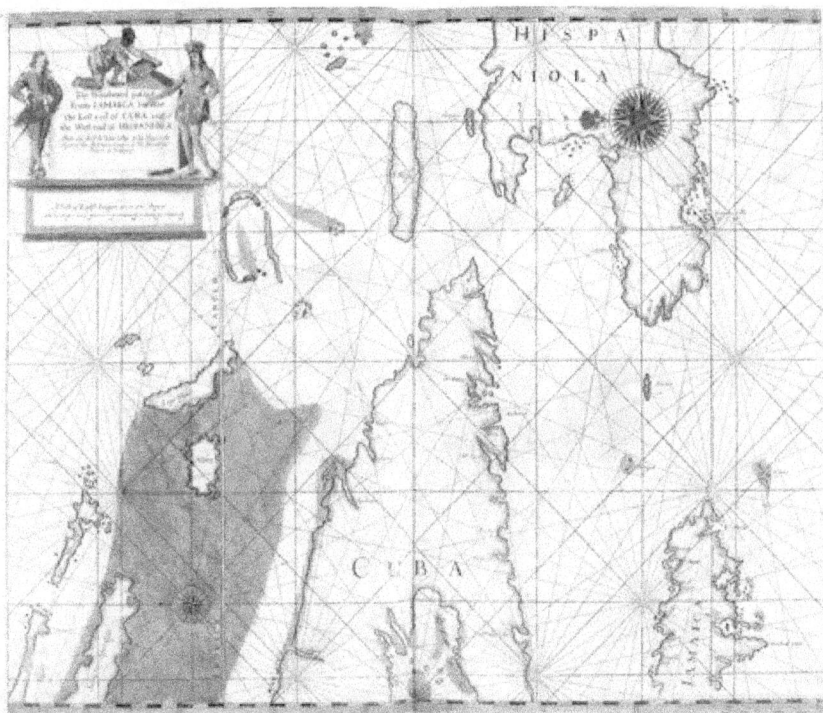

amaü

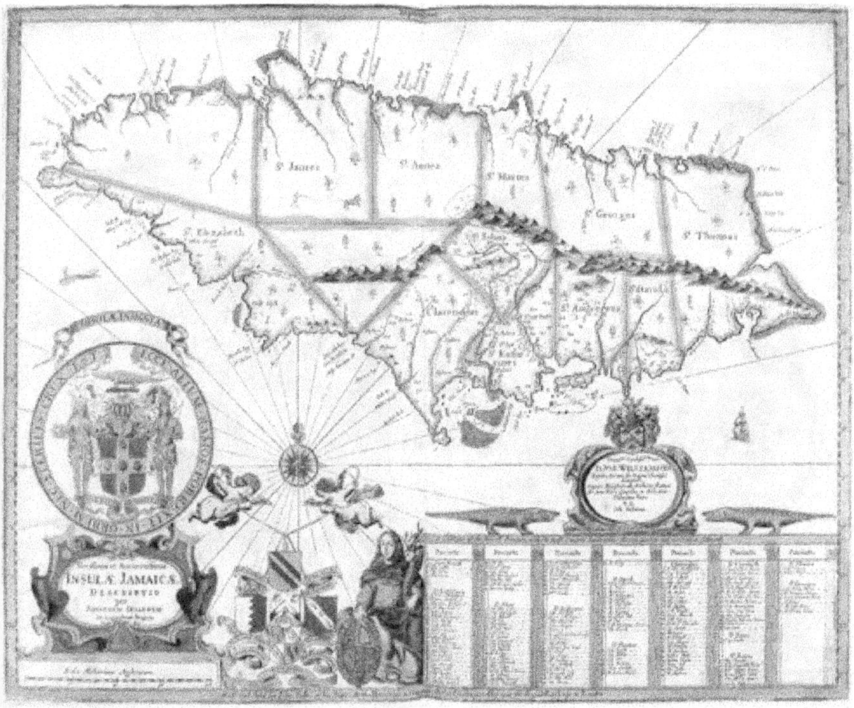

INSULÆ JAMAICÆ
DESCRIPTIO
St. James
St. Anns
St. Marys
St. Georges
St. Thomas
St. Elizabeth
Clarendon

Chart of the Sea coast of Brazil

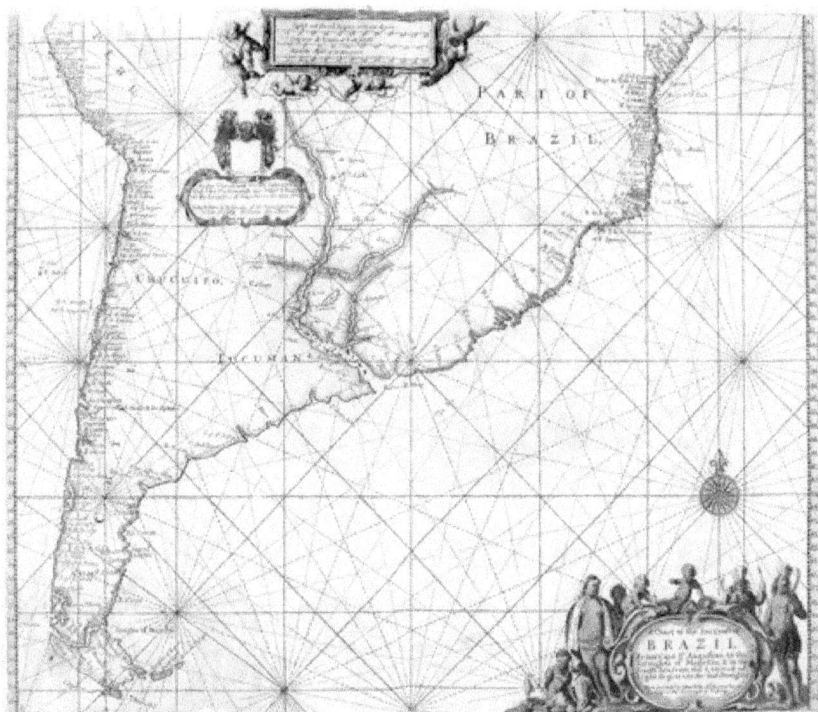

CHICUITO.

TUCUMAN

A Chart of the South
BRAZIL.

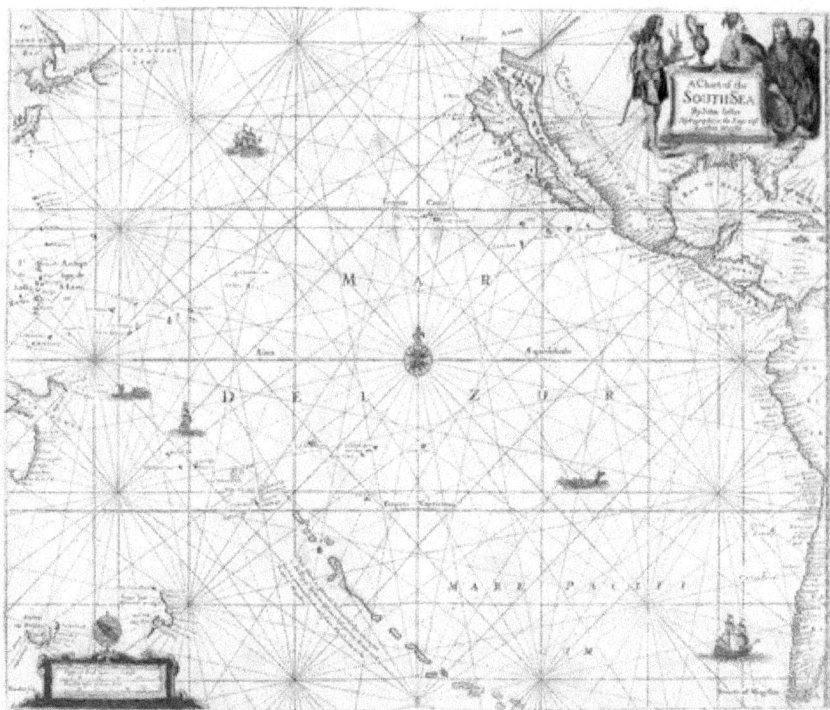

A Chart of the SOUTH SEA By John Seller Hydrographer to the King &c.